Technician's Guide to
Programmable Controllers

SECOND EDITION

RICHARD A. COX
SPOKANE COMMUNITY COLLEGE

Delmar Publishers Inc.®

NOTICE TO THE READER

Cover: Programmable controller photos courtesy of: Allen-Bradley Company, Highland Heights, OH
Gould Inc., Andover, MA
Square D Company, Milwaukee, WI, Richard Brodzeller, photographer

Delmar Staff
Associate Editor: Cameron O. Anderson
Editing Manager: Barbara A. Christie
Project Editor: Lawrence T. Main

For information, address Delmar Publishers, Inc.
2 Computer Drive West, Box 15-015
Albany, New York 12212

10 9 8 7 6 5 4 3 2

Printed in the United States of America
Published simultaneously in Canada
by Nelson Canada,
A Division of International Thomson Limited

Library of Congress Cataloging-in-Publication Data

Cox, Richard A.
  Technician's guide to programmable controllers / Richard A. Cox.–2nd ed.

    p.   cm.
  Includes index.
  ISBN 0-8273-2830-3 (pbk.)
  1. Programmable controllers.   I. Title.
TJ223.P76C69 1989
629.8'95--dc19

88-11911
CIP

Contents

CHAPTER 10

CHAPTER 11

CHAPTER 12

CHAPTER 13

CHAPTER 14

CHAPTER 15

CHAPTER 16

CHAPTER 17

Preface

The programmable controller, which was first introduced in 1969, has become an unqualified success. Programmable controllers, or **PC's** as they are often referred to, are now produced by over 50 manufacturers. Varying in size and sophistication, these electronic marvels are rapidly replacing the hard-wired circuits that have controlled the process machines and driven equipment of industry in the past.

With every major motor control equipment manufacturer and some computer companies now offering PC's, it would be impossible to write a book that explained how they all work and/or are programmed. Instead, this book is intended to discuss PC's in a general or generic sense, and to cover the basic concepts of operation that are common to all PC's.

Many technicians seem apprehensive about PC's and their application in industry. One of the purposes of this text is to explain PC basics in a plain, easy to understand approach so that technicians with no PC experience will be more comfortable with their first exposure to a programmable controller.

Half the battle of understanding any programmable controller is to first understand the terminology of the PC field. This text covers terminology as well as explaining the input/output section, processor unit, programming devices, memory organization, and more.

A chapter on understanding and using ladder diagrams has been included to explain not only ladder diagrams, but also relay ladder logic which is the language of most programmable controllers.

Examples of basic programming techniques with typical PC's are discussed and illustrated, as well as some data manipulation instructions. While the text can only scratch the surface of data manipulation and other advanced programming capabilities of the PC's on the market today, the reader will gain a basic understanding of data moves and how they work. Like any new skill, a firm base of understanding is required before a technician can become proficient at that skill. After completing the text, the reader will possess a good foundation upon which additional PC skills and understanding can be built.

The best teacher, of course, is experience, and as stated several times in the text, the only way to really understand any given PC is to work with that PC. If a PC is not readily available, the next best thing is a workshop or seminar sponsored by a local PC distributor. If a workshop or seminar is not available, then obtain as much literature and other information as possible from a local electrical distributor or PC representative.

The PC manufacturers are reducing prices while adding new features and program capabilities every day. They are enjoying a growth rate paralleled only by the home computer field. Even with the rapid advancements in PC's, the technician without an electronics background need not feel intimidated. The PC manufacturers are doing everything possible to make the PC's easy to install, program, troubleshoot, and maintain.

The author wishes to acknowledge the cooperation and assistance of the different manufacturers whose product information and photographs are used throughout the text to illustrate different concepts and PC components. All these manufacturers are leaders in the PC field, and each offers a full line of programmable controllers. It is not a question of which PC is best, but rather which PC best fits your application.

Richard A. Cox is the Department Chairman of the Electrical/Robotics Department at Spokane Community College in Spokane, Washington and is also a member of the International Brotherhood of Electrical Workers Local 73.

Chapter 1

What is a Programmable Controller (PC)?

Objectives

After completing this chapter, you should have the knowledge to
- Describe several advantages of a programmable controller over hard wired relay systems.
- Identify the three main components of a typical programmable controller and describe the function of each.
- Define the term *discrete*.
- Identify different types of programming devices.

A programmable controller is a solid-state device designed to perform the logic functions previously accomplished by electro-mechanical relays, drum switches, mechanical timers/counters, etc., for the control and operation of manufacturing process equipment and machinery.

Even though the electro-mechanical relay (control relay) has served well for many generations, often under adverse conditions, the ever increasing sophistication and complexity of modern processing equipment requires faster acting and more reliable control than electro-mechanical relays and/or timing devices can offer. Relays have to be hard wired to perform a specific function, and when the system requirements change, the relay wiring has to be changed or modified. In extreme cases, such as in the auto industry, complete control panels had to be replaced since it was not economically feasible to rewire the old panels with each model changeover.

It was, in fact, the requirements of the auto industry and other highly specialized, high speed manufacturing processes that created a demand for smaller, faster acting, and more reliable control devices. The electrical/electronics industry responded with modular designed solid-state electronic devices.

These early devices, while they offered solid-state reliability, lower power consumption, expandability, and elimination of much of the "hard-wiring," also brought with them a new language. The language consisted of AND gates, NOR gates, NOT gates, OFF RETURN MEMORY, J-K flip flops, and so on.

1

What happened to simple relay logic and ladder diagrams? That is the question the plant engineers and maintenance electricians were asking the solid-state device manufacturers. The reluctance of the end user to learn a new language and the advent of the microprocessor gave the industry what is now known as the **Programmable Controller** (PC). Internally there are still AND gates, OR gates, and so forth in the processor, but the design engineers have pre-programmed the PC so that circuits can be entered using RELAY LADDER LOGIC. While RELAY LADDER LOGIC may not have the mystique of other computer languages such as FORTRAN and COBOL, it is a high level, real world, graphic display that is understood by all electricians.

What does a PC consist of, and how is it different from a computer control system? The PC consists of a processor unit and an **interface** like a computer system (Figure 1-1), and while there are similarities, there are some very major differences.

NOTE: An interface is where two systems come together and interact or communicate. In the case of a PC, the communication or interaction is between the inputs/outputs and the processor.

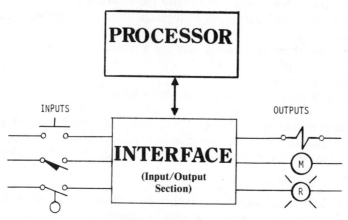

Figure 1-1. Processor Unit and Interface

The PC is designed to operate in the industrial environment with wide ranges of ambient temperature and humidity and is not usually affected by the electrical noise that is inherent in most industrial locations.

NOTE: Electrical noise is discussed in Chapter 2.

PC's are designed to be operated by plant engineers and maintenance personnel without knowledge of computers or computer technology. Like the computer, which has an internal memory for the storage of a program, the PC also has a memory for storing the user program, or LOGIC, for controlling the operation of a process machine or driven equipment. But unlike the computer, the PC is programmed in RELAY LADDER LOGIC, not one of the computer languages. It should be stated, however, that some PC's can be programmed using Boolean Algebra as well as relay ladder logic. A brief description of Boolean Algebra will be covered in Chapter 9.

2

Maybe one of the biggest, or at least most significant, differences between the PC and a computer is the fact that PC's have been designed for installation and maintenance by plant electricians who are not required to be highly skilled electronics technicians. Troubleshooting is simplified by the design of most PC's in that they include fault indicators, and written fault information that is displayed on the programmer CRT (Cathode Ray Tube), which is similar to a TV screen.

A typical PC can be divided into three components as shown in Figure 1-2. These components consist of the **processor unit,** the **input/output section** (interface), and the **programming device.**

processor unit

input/output section

programming device

Figure 1-2. PC Components
(Courtesy of Westinghouse Electric)

The **processor unit** houses the processor which is the "brain" of the system. This brain is a microprocessor-based system which replaces control relays, counters, timers, sequencers, and so forth and is designed so the user can enter the desired circuit in relay ladder logic. The processor then makes all the decisions necessary to carry out the user program for control of a machine or process. It can also perform arithmetic functions, data manipulation, and communications between the PC, remotely located PC's, and/or computer systems. A DC power supply is required to produce the low level DC voltage used by the processor. This power supply can be housed in the processor unit or may be a separately mounted unit depending on the model and/or the manufacturer.

NOTE: Some manufacturers refer to the processor as a **CPU** or central processing unit.

The **input/output section** consists of input modules and output modules. (The number of each is determined by the process equipment and/or driven machine requirements.) Input and output modules, referred to as the I/O (I for input and O for output) are where the "real world" devices are connected. The "real world" input (I) devices might be pushbuttons, limit switches, analog sensors, thumbwheels, selector switches, and so on, while the "real world" outputs (O) could be hard wired to motor starters, solenoid valves, indicator

3

lights, position valves, and the like. Actual devices are also referred to as **discrete** inputs and **discrete** outputs. The terms "real world" or "discrete" are used to separate actual devices that exist and must be physically wired as compared to the user program and logic of the system that duplicates the function of relays, timers, counters, and so on even though none actually exists. This may seem a bit strange and hard to understand at this point, but the distinction between what the processor can do internally, which eliminates the need for all the previously used control relays, timers, counters, and so forth, will be graphically shown and readily understandable further in the text.

A reference was made earlier in this chapter to the I/O section as an interface. Though not a common reference, it is an accurate one. The I/O section contains the circuitry necessary to convert input voltages 120–240V AC or DC from input devices to low level DC voltages (signals) that the processor uses. Likewise, the output module changes low level DC signals from the processor to 120–240V AC or DC voltages required to operate the output devices. This is a brief overview of the I/O section and its function. How input and output devices are wired to I/O modules and more about the module circuitry itself is covered in Chapter 2.

The **programming device** may be called an **industrial terminal** (Allen-Bradley), **program development terminal** (General Electric), **programming panel** (Modicon), or simply **programmer** (Square D). Regardless of their names, they all perform the same function and are used to enter the desired program in relay ladder logic that will determine the sequence of operation and ultimate control of the process equipment or driven machinery. The programming device may be a hand held unit with an **LED** (light emitting diode) display, an **LCD** (liquid crystal display) as shown in Figure 1-3, a desktop type with a CRT display (Figure 1-4) or other compatible computer terminals. Figure 1-5 shows an IBM PC laptop computer that can be used with the Westinghouse PC 1100. All major PC manufacturers now offer software which will allow programming with an IBM or IBM compatible.*

*IBM is a registered trademark of the IBM Corporation.

Figure 1-3. Hand Held Programmer
(Courtesy of Gould, Inc.)

Figure 1-4. Desktop Programmer
(Courtesy of Allen-Bradley)

Figure 1-5. Programmable Controller and Computer used as a programming device
(Courtesy of Westinghouse Electric)

As we will find out later when we discuss programming techniques, there is as much variation in programming as there is in the names for the programming devices.

Chapter Summary

Programmable controllers have made it possible to precisely control large process machines and driven equipment with less physical wiring and wiring time than is required with standard electro-mechanical relays, pneumatic timers, drum switches, and so on. The programmability allows for fast and easy changes in the relay ladder logic to meet the changing needs of the process or driven equipment without the need for expensive and time consuming rewiring. By designing the modern PC to be "electrician friendly," the PC can be programmed and used by plant engineers and maintenance electricians without electronic backgrounds.

Review Questions

1. Name the three main components of a typical programmable controller (PC).
2. What does the abbreviation I/O mean?
3. "Real world" or actual input or output devices are also referred to as _____ devices.
4. List two types of programming devices.
5. T   F   The power supply needed to produce the low level DC voltage used by the processor is always located inside the processor unit.
6. What do the following initials or acronyms stand for?

LED   I/O

LCD   PC

CPU

Chapter 2

Understanding the Input/Output (I/O) Section

Objectives

After completing this chapter, you should have the knowledge to
- Describe the I/O section of a programmable controller.
- Identify DIP switches.
- Describe how basic AC/DC input and output modules work.
- Define *optical isolation* and describe why it is used.
- Describe the proper voltage connections used for both input and output modules.
- Explain why a hard wired emergency stop function is desirable.
- Define the term *interposing*.
- Describe what I/O shielding does.
- Describe surge suppression.

The I/O section consists of an I/O rack (Figure 2-1) and individual I/O modules (Figure 2-2). The I/O section may also include a DC power supply (Figure 2-1, Item 5) when a separate power supply is used in addition to, or in conjunction with, the DC power supply usually found in the processor unit.

1. 7 Position DIP Switch (10 Per Rack)
2. 41-Pin Connector (11 Per Rack)
3. Logic Power On/Off Circuit Breaker
4. Power On Indicator
5. I/O Power Supply
6. Terminal Board
   115V/230V AC Input Selected
   By Jumper
7. Tray For Containing Field Wiring
8. Cardguide (11 Per Rack)

Figure 2-1. I/O Rack
(Courtesy of General Electric)

6

1. Input "On" Indicators
2. Markable Lens Surface
3. Box Lug Terminals
4. Cover

Figure 2-2. Typical Input Module
(Courtesy of General Electric)

The I/O rack can also be part of one main rack that houses not only the I/O modules but also the processor like the Square D SY/MAX Model 300 shown in Figure 2-3.

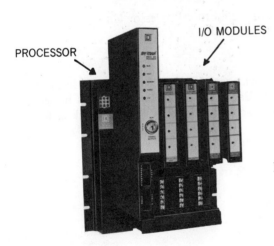

PROCESSOR

I/O MODULES

Figure 2-3. Square D SY/MAX Model 300
I/O Rack and Processor
(Courtesy of Square D Company)

I/O RACK

Mounted on the backplane of the I/O rack are pre-wired slots or connectors (Figure 2-1, Item 2) into which the individual I/O modules are inserted. Proper alignment when inserting the modules is assured by card guides (also referred to as printed circuit board guides) which are mounted on the top and bottom of the I/O rack (Figure 2-1, Item 8).

Normally, more than one (1) I/O rack can be connected to a processor unit via interconnecting cables (Figure 2-4).

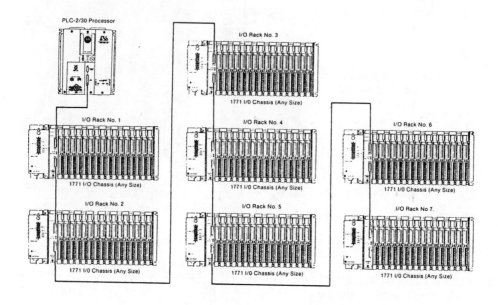

Figure 2-4. Seven I/O Racks Connected to a Processor Unit
(Courtesy of Allen-Bradley)

To allow the processor to communicate properly with each I/O rack, **DIP** (dual-in-line package) switches have been mounted in each rack. Each of the DIP switches must be set in the proper sequence as prescribed by the manufacturer to identify Rack 1, Rack 2, Rack 3, etc., (Figure 2-1, Item 1, and Figure 2-5).

SWITCH
GROUP
ASSEMBLY

POWER
SUPPLY
SOCKET

BACKPLANE
CONNECTOR

Figure 2-5. I/O Rack DIP switches
(Courtesy of Allen-Bradley)

8

To identify the I/O rack below the Allen-Bradley processor in Figure 2-4 as rack one (1), switches 3, 4, and 5 would be turned ON, as indicated in the product literature, Figure 2-6. To identify the next I/O rack in Figure 2-4 as rack two (2), the DIP switches mounted in that rack would be set as follows: Switch 3 and 4 ON, Switch 5 OFF (Figure 2-6).

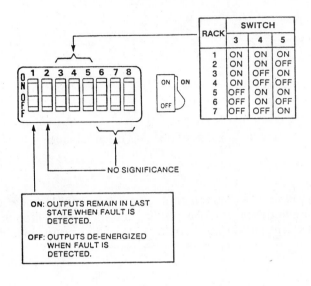

RACK	SWITCH		
	3	4	5
1	ON	ON	ON
2	ON	ON	OFF
3	ON	OFF	ON
4	ON	OFF	OFF
5	OFF	ON	ON
6	OFF	ON	OFF
7	OFF	OFF	ON

— NO SIGNIFICANCE

ON: OUTPUTS REMAIN IN LAST STATE WHEN FAULT IS DETECTED.

OFF: OUTPUTS DE-ENERGIZED WHEN FAULT IS DETECTED.

Figure 2-6. Configuration of DIP switches
(Courtesy of Allen-Bradley)

Whether only one I/O rack is used, or several are used, all enclosures including the processor unit must be grounded according to the manufacturer's specifications and any applicable local, state, and National Electrical Codes.

The rack number, location of a module within a rack, and the terminal number of a module to which an input or output device is connected will determine the device's **address.** Each input and output device must have a distinct **address** so the processor knows where the device is located and in return can send and receive signals, enabling the processor to monitor and/or control the device. Addresses and addressing of input and output devices will be covered in Chapter 4.

AC/DC DIGITAL INPUT MODULES

The input module provides the interfacing between the ON/OFF voltage condition of the actual "real world" or **discrete** input devices such as limit switches and pushbuttons, and the logic or low level DC voltage required by the processor. Stated more simply, the input module communicates the status of the input devices, whether they are closed (ON) or open (OFF), to the processor. One input module can receive inputs from 4, 6, 8, 10, 16, or 32 individual input devices depending on the manufacturer. Figure 2-7 shows a simplified diagram for one input circuit of a typical AC input module. A bridge rectifier converts the AC input (120–240V) to DC and resistors are used to drop the voltage to logic level. The actual value of the low level DC logic voltage may vary with different manufacturers but is usually 5–12 volts.

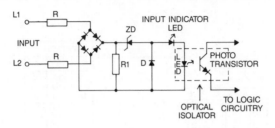

Figure 2-7. Simplified AC Input Circuit
with Optical Isolation

Figure 2-8 depicts a typical DC input module circuit. Since this module has a DC input, there is no need for a bridge rectifier, but resistors are still used to drop or scale down the DC input voltage (which can be from 12 to 240V depending on the module) to logic level (5 to 12V).

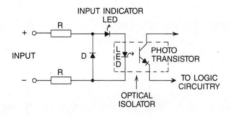

Figure 2-8. Simplified DC Input Circuit
with Optical Isolation

Also notice in Figures 2-7 and 2-8 that both the AC and DC input module circuits employ **optical isolation.** The optical isolation not only separates the line or input voltage from the logic circuits but also prevents voltage transients from damaging the processor. Optical isolation also helps reduce the effects of electrical noise, common in the industrial environment, which can cause erratic operation of the processor. Optical isolation is achieved by using an LED and a **photo-transistor** that acts as a receiver as shown in Figures 2-7 and 2-8. When an input device is closed (ON), the LED is also turned ON. The light emitted from the LED is picked up by the photo-transistor, and the status of the input device is communicated in logic or low level DC voltage to the processor. Coupling and isolation could also be accomplished by using a pulse transformer.

An additional LED is used with each circuit of an input module to indicate the status of the input device as shown in Figures 2-7 and 2-8. The LED will be ON when the input device is closed or ON. The input indicator LED is used as a troubleshooting aid and will be discussed later in Chapter 17.

A typical I/O module consists of two parts: a printed circuit board and a terminal assembly. The printed circuit board plugs into a slot or connector in the I/O rack and contains the solid-state electronic circuits that interface the input and/or output devices with the processor. The terminal assembly then attaches to the front edge of the printed circuit board, which may or may not have a protective cover, depending on the manufacturer.

Figures 2-9a, b, and c show typical input modules. Note that the Westinghouse input module can be ordered with pictorial lenses which are illuminated by an LED behind the lens when the input device is closed or ON.

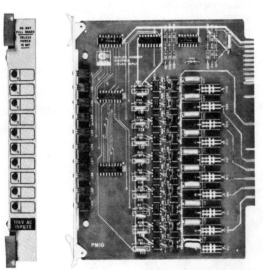

Figure 2-9a.
(Courtesy of Cutler-Hammer)

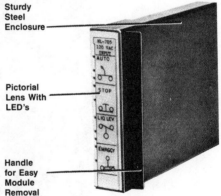

Figure 2-9b.
(Courtesy of Westinghouse Electric)

Sturdy
Steel
Enclosure

Pictorial
Lens With
LED's

Handle
for Easy
Module
Removal

Figure 2-9c.
(Courtesy of Allen-Bradley)

1. Red identification label
2. Status indicators
3. Protective covers
4. Field Wiring Arm connects here
5. Labels identify user inputs
6. Slotted for I/O slot insertion only

Once the Allen-Bradley input module (Figure 2-9c) is installed in the I/O rack (Figure 2-10), a field wiring arm (terminal assembly) is swung up and locked in place (Figure 2-11).

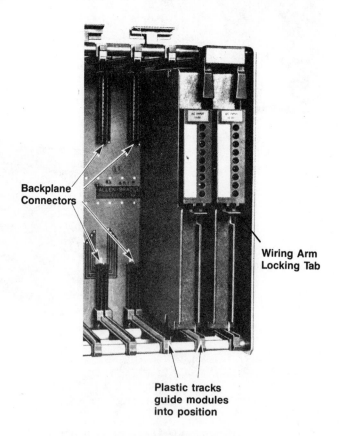

Figure 2-10. Input Module installed in I/O Rack
(Courtesy of Allen-Bradley)

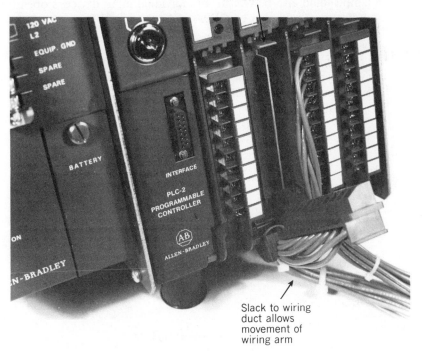

Module lever holds wiring Arm when in Position

Slack to wiring duct allows movement of wiring arm

Figure 2-11. I/O Module field wiring arm
(Courtesy of Allen-Bradley)

After the input modules have been installed in the I/O rack, they are ready to have the discrete input devices connected to their terminals. Figure 2-12 shows input devices and the proper AC line connections for an eight (8) circuit input module. The L1 lead would be the HOT (HI) for an AC module and would be POS (+) for a DC module; likewise, the L2 wire (terminal B) would be the neutral (LOW) for an AC module and would be the NEG (−) lead for a DC module.

Figure 2-12. Typical wiring diagram for an 8-circuit Input Module

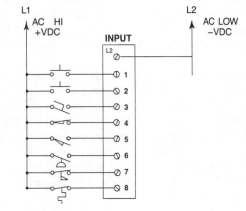

13

Input modules will have LED or neon indicators for each input to indicate the state of the input device, either ON or OFF. When an individual input device is closed (ON), the LED or neon lamp should also be ON. If an individual input device is open (OFF), the LED or neon indicator should also be OFF.

AC OUTPUT MODULES

Figure 2-13 shows a typical AC output module circuit. The output usually consists of a **Triac** (as shown), but in some cases an SCR may be used.

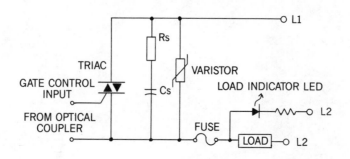

Figure 2-13. Simplified AC
Output Circuit

The Triac is the equivalent of two **SCR's** (silicon-controlled rectifiers) in inverse parallel connection with a common **gate** that controls the switching state. Once the breakover voltage point is reached on the gate (normally 1 to 3 volts), the Triac conducts in either direction. The gate pulse or signal which turns on the Triac, which, in turn, turns on an output device, is controlled by a logic level signal from the processor. This logic level signal is optically isolated to prevent a failure in the output module from damaging the processor.

A Triac is a solid block of crystalline material and is more sensitive to applied voltages and currents than standard relay contacts. Triacs are also limited to a maximum peak applied voltage, and, if this value is exceeded, a "dielectric type" breakdown can result, causing a permanent short-circuit condition. Peak voltages higher than normal can be the result of transients and noise on the output line. To limit transients and noise, a varistor (THY) as shown in Figure 2-13 is placed across the Triac.

Triacs being constructed of a solid "block of material" have some characteristics that are not found with standard relay contacts. The Triac, rather than having ON and OFF states, actually has LOW and HIGH resistance levels respectively. In its OFF-state (high resistance), a small leakage current still flows through the Triac. This leakage current, which is usually only a few milliamperes or less, normally will cause no problem. When low resistive pilot lights are connected to AC output modules a faint glow of the filament may be detectable, even when the module is off. Likewise, the coils of some small control relays and/or solenoids may produce a detectable hum due to the Triac leakage current even though the Triac is technically off.

Individual AC outputs are given a load current rating, usually 2 to 3 amps, and a maximum load for any one module is also given. Along with the load current ratings, there will also be surge or inrush current ratings. While Triacs are capable of carrying surge currents higher than their continuous current rating, such surges must be of short duration (1/2 to one cycle) and not repetitive. Exceeding the manufacturer's listed surge values or the maximum continuous current rating, usually referred to as maximum RMS on-state current, will result in a permanent short.

After an output module is installed in the I/O rack, the actual "real world" or discrete output devices are connected. Figure 2-14 shows the proper termination for an eight (8) circuit AC output module.

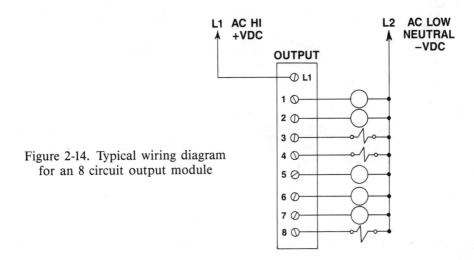

Figure 2-14. Typical wiring diagram for an 8 circuit output module

SAFETY CIRCUIT

The National Electrical Manufacturing Association (NEMA) standards for programmable controllers recommends that consideration be given to the use of emergency stop functions which are independent of the programmable controller. The standard reads in part: "When the operator is exposed to the machinery, such as loading or unloading a machine tool, or where the machine cycles automatically, consideration should be given to the use of an electro-mechanical override or other redundant means, independent of the controller, for starting or interrupting the cycle."

While today's PC's are very rugged and dependable, where safety is concerned **DON'T** depend on the solid-state devices and circuitry of the PC. The NEMA recommendation recognizes the importance of a hard wired stop circuit (as shown in Figure 2-15) to remove power to the output devices. A second set of MCR contacts could be added in the X1 line to remove power to the input devices. It is common, however, to put the MCR contacts on the output side only so that the inputs can remain energized for troubleshooting.

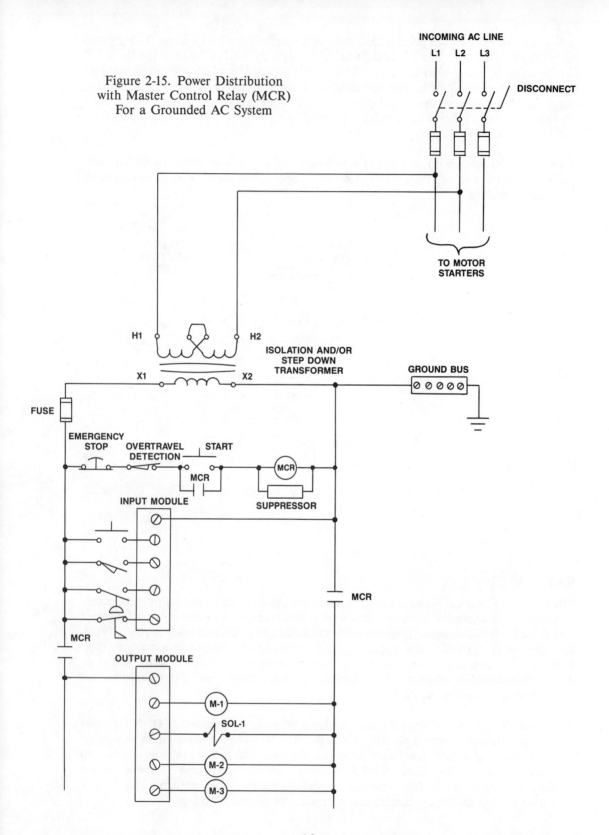

Figure 2-15. Power Distribution with Master Control Relay (MCR) For a Grounded AC System

16

It is also worth noting that solid-state output devices usually (though not absolutely) fail shorted, rather than in an OPEN condition. By failing in a shorted or ON condition an added safety hazard is possible if a hard wired E-STOP or master control is not included as part of the PC installation.

To protect the output module circuits, neither the load nor inrush current levels should be exceeded. On most output modules the outputs are individually fused, but do not rely on the fuses alone.

NOTE: The fuses that are used in output modules have been carefully selected by each manufacturer for current/time characteristics, and only the fuses recommended in the product literature should be used.

As a troubleshooting aid, output modules are equipped with a single LED or neon lamp that turns on if a fuse has been blown or individual blown fuse lights for each output as indicated in Figure 2-16, Item 5. Additionally, individual LED or neon lamps are used for each output to indicate when the output circuit is ON (Figure 2-16, Item 4). The output circuit LED indicates only that the output circuitry has been turned on. It is not an indication that current is actually flowing through the output device.

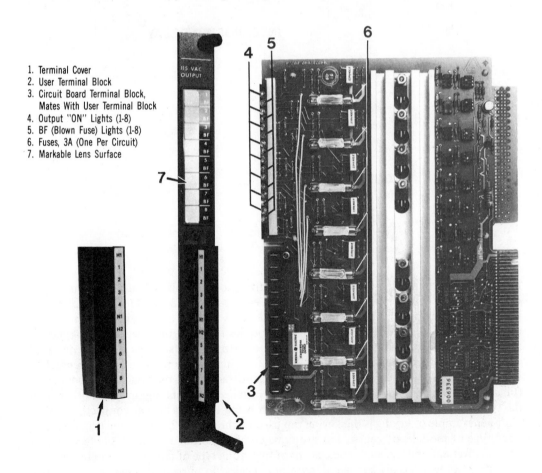

1. Terminal Cover
2. User Terminal Block
3. Circuit Board Terminal Block, Mates With User Terminal Block
4. Output "ON" Lights (1-8)
5. BF (Blown Fuse) Lights (1-8)
6. Fuses, 3A (One Per Circuit)
7. Markable Lens Surface

Figure 2-16. General Electric 115V AC Output Module

17

NOTE: For controlling loads larger than the rating of an individual output circuit, a standard control relay, which has a small inrush and sealed current value, is connected to the output module. The contacts of the control relay, which are rated at 10 amps or more, can then be used to control a larger load. Depending on the rating of the output module, this is a common practice for NEMA size 4 and larger motor starters. When a control relay is used in this manner, it is called an **interposing relay** (Figure 2-17).

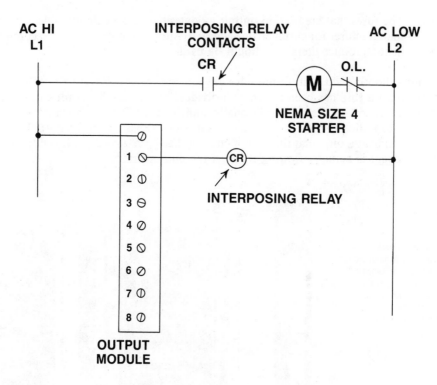

Figure 2-17. Interposing Relay

Figure 2-18 shows an Allen-Bradley 240V AC output module. This AC output module looks just like the AC or DC input module discussed earlier (Figure 2-9c). This is not only true for Allen-Bradley but also for other manufacturers as well. In all instances the manufacturers label and color code the fronts of all modules to distinguish between the types of modules (AC input, DC output, TTL, Analog, etc.). Most manufacturers also have designed each module so it can be **keyed.** Notice Figure 2-18, Item 6; the module has been notched in two places. Installing keying bands on the Allen-Bradley I/O rack backplane connector (Figure 2-19) where a specific module is to be installed prevents any module, other than the type for which the connector is keyed, from being installed in that connector or slot. Another example of keying is the keying pin used by the Square D Company (Figure 2-20). Each type of module, of course, has a unique combination of notches. This feature prevents inadvertent or accidental replacement of the wrong type of module, say an input module, into a slot which is already wired to output devices. To prevent damage and down time it is **important** that the keying system be used.

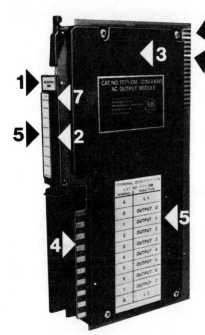

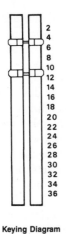

Figure 2-18. Notched AC Output Module
(Courtesy of Allen-Bradley)

1. Orange identification label
2. Status indicators
3. Protective cover
4. Field Wiring Arm connects here
5. Labels identify user outputs
6. Slotted for I/O slot insertion only
7. Blown fuse indicator

Figure 2-19. Backplane Connector Keying Diagram
(Courtesy of Allen-Bradley)

Keying Diagram

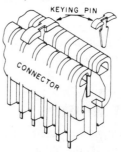

Figure 2-20. Connector Keying Pin
(Courtesy of Square D Company)

DC OUTPUT MODULES

DC output modules are available in ranges from 12–240V DC depending on the manufacturer. While the DC module uses a power transistor rather than a Triac, as the AC output module does, the power transistor is also sensitive to excessive applied voltages and surge currents. Be sure and check the specifications of the output module being installed to ensure that the applied voltage and continuous current ratings are not exceeded.

TRANSISTOR-TRANSISTOR LOGIC (TTL) INPUT AND OUTPUT MODULES

TTL input modules are designed to be compatible with other solid-state controls, sensing instruments, many types of photoelectric sensors, and some 5V DC level control devices. TTL output modules are used for interfacing with discrete or integrated circuit (IC) TTL devices, LED displays, and various other 5V DC devices.

ANALOG INPUT AND OUTPUT MODULES

Analog input modules are used to convert analog signals from analog devices that sense such variables as temperature, light intensity, speed, pressure, and position to 12-Bit Binary or to three digit Binary-Coded Decimal (BCD), depending on the manufacturer, for use by the processor. The Analog output module changes the 12-Bit Binary or 3-Digit BCD value used by the processor into analog signals. These analog signals could be used for speed controllers, signal amplifiers, or valve positioners. Binary and Binary Coded Decimal (BCD) will be covered in Chapter 4.

THERMOCOUPLE INPUT MODULE

As the name implies, this module is used to convert thermocouple and other millivolt signals to Binary or Binary-Coded Decimal (BCD) values for use by the processor.

REED RELAY OUTPUT MODULE

The reed relay type output module is used where dry reed relays are desirable. They may be used for low-level switching (small current-low voltage), or, depending on the manufacturer, they may be used for switching 250V AC/DC at 2 amps. Reed relay output modules are cheaper than normal solid-state AC/DC output modules and work well, and are usually recommended, when indicator lamps are the output devices. Reed relay modules are available with normally open (N.O.) contacts, normally closed (N.C.) contacts, or a combination of both N.O. and N.C. contacts, again depending on the manufacturer.

NOTE: PC manufacturers are introducing new modules and special application modules almost daily. A few basic modules have been discussed in this chapter for a basic understanding only. Contact your local PC representative(s) for a complete list of modules that are available.

ELECTRICAL NOISE

Electrical noise can be generated whenever inductive loads such as relays, solenoids, motor starters and motors are operated by "hard contacts" such as pushbuttons, selector switches, relay contacts, etc. The noise, or high transient voltages are caused by the collapsing magnetic field when an inductive device is switched OFF.

NOTE: The term hard contact is used to differentiate between solid-state switching devices and standard mechanical contacts.

To minimize noise and reduce the chance of temporary operating errors the following steps should be considered.

1. Mount equipment in metal enclosures when possible, because metal helps protect against electromagnetic radiation.
2. Connect the processor and I/O to an AC power source separate from the AC source of any potential noise generating equipment.
3. Run the AC lines to the PC and I/O in separate conduits.
4. Use an isolation transformer to supply AC power to the processor and I/O when normal AC supply is very noisy (Figure 2-15).
5. Add interference and surge suppression devices as required.
6. Ground all equipment and shielded conductors in accordance with the manufacturer's recommendations and/or local codes.

Figure 2-21 illustrates an input configuration that would require surge suppression to reduced interference from an inductive load.

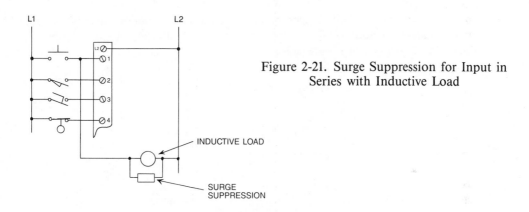

Figure 2-21. Surge Suppression for Input in Series with Inductive Load

Figure 2-22 is an example of an output device in series with a pushbutton switch. This installation could generate electrical noise and would require surge suppression.

Figure 2-22. Surge Suppression for Switch in Series with Inductive Load

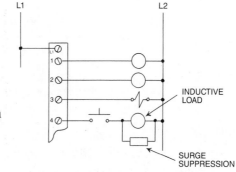

Figure 2-23 shows another method of wiring an output device with a switch in parallel that should have surge suppression.

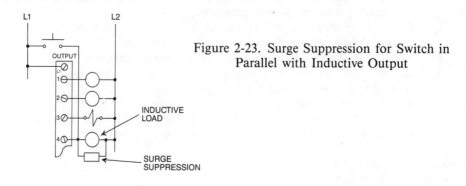

Figure 2-23. Surge Suppression for Switch in Parallel with Inductive Output

The type of surge suppressor will depend on the size and type of load. Consult your equipment representative for additional information. Some typical types of surge suppression for AC devices are shown in Figure 2-24.

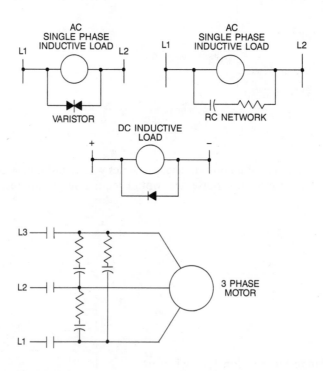

Figure 2-24. Typical Connections for Suppression of Small Inductive Loads

22

With solid-state control systems, proper grounding helps eliminate the effects of electro-magnetic induction (EMI). Figure 2-25 shows a typical installation using an equipment grounding conductor to connect several PC's and/or I/O racks together. The equipment grounding conductor is attached to the metal frame of the PC and/or I/O rack with a ground lug. A detail of the connection is shown in Figure 2-26.

NOTE: Check local codes and the manufacturer's specifications to ensure proper installation.

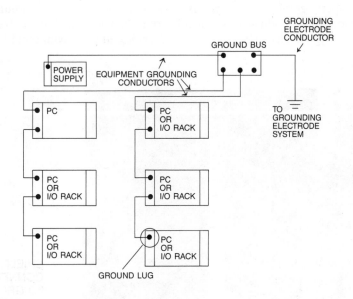

Figure 2-25. Typical Equipment Grounding Configuration

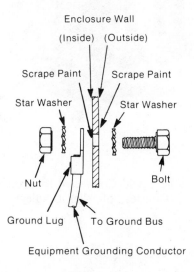

Figure 2-26. Detail of Grounding Lug Attachment to
PC and/or I/O Frame (Courtesy of Allen-Bradley)

I/O SHIELDING

Certain I/O modules (TTL, analog, thermocouple, etc.) will require shielded cable to reduce the effects of electrical noise. The cable shield, which surrounds the cable conductors, shields the conductors from electrical noise.

When installing shielding cable it is important that the shield only be grounded at one end. If the shield was grounded at both ends a ground loop would be created and the ground loop could introduce ground currents which could result in faulty operation of the processor.

Because a properly grounded I/O rack is already connected to earth ground through an equipment grounding conductor, the shield should be terminated at the I/O rack, not at the device end. Figure 2-27 shows the shield of a shielded cable connected to the I/O rack frame.

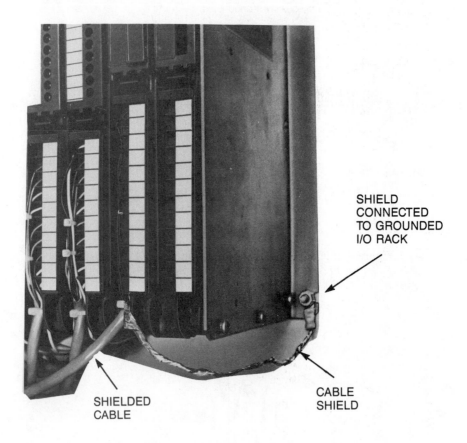

Figure 2-27. Cable Shield Connected to Grounded I/O Rack
(Courtesy of Allen-Bradley)

Figure 2-28 shows a shielded/twisted pair cable connected to a sensing device and I/O rack. Note that the shield is connected at one end only.

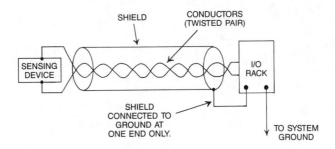

Figure 2-28. Shield Connected to Ground at One End Only

Chapter Summary

The I/O rack houses the individual I/O modules that are connected to "real world" or discrete devices. The input modules act as an interface so the processor can communicate with the inputs and control the outputs. The I/O rack connects to the processor unit via an interconnecting cable(s), through a bus duct (Figure 2-29), or the I/O may be in the same rack as the processor (Figure 2-30). Consideration must be given to surge suppression and I/O shielding if proper operation is to be ensured.

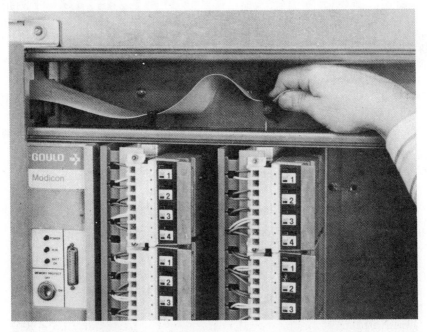

Figure 2-29. Interconnecting Cable from Processor Plugs
into I/O Housing (Courtesy of Gould Inc.)

Figure 2-30. Power Supply, Processor (Controller) and I/O
Modules Mounted on Same Rack (Chassis)
(Courtesy of Gould, Inc.)

Review Questions

1. Describe briefly the purpose of the I/O section.
2. State two reasons for employing optical isolation.
3. Draw an input module with four input devices and show all necessary electrical connections and identify potentials (L1-L2).
4. Draw an output module with four output devices and show all necessary electrical connections and identify potentials (L1-L2).
5. T F Triacs are susceptible to "dielectric type" breakdown if the maximum peak voltage level is exceeded.
6. Briefly describe why a hard wired emergency stop circuit is recommended for programmable controller installations.
7. Briefly describe the function of an interposing relay.
8. T F I/O modules are keyed to prevent unauthorized personnel from removing them from the I/O rack.
9. Which of the following are **NOT** normally a source of electrical noise.
   a. Solenoid
   b. Relay
   c. Indicator lamp
   d. Motor starter
   e. Motor
   f. Overload heaters
10. T F To ensure maximum benefit of shielding, the shield of a shielded cable must be terminated and grounded at both ends.

11. E-STOP refers to:
    a. Extra stop
    b. Emergency stop
    c. Every stop
    d. Elevator stop
    e. Energy stop
12. T F Electro-magnetic induction (EMI) can be reduced with proper grounding of equipment.
13. Solid-state output devices tend to:
    a. Never fail
    b. Fail in the open (OFF) condition
    c. Fail in the shorted (ON) condition
    d. Not be affected by overload

Chapter 3

Processor Unit

Objectives

After completing this chapter, you should have the knowledge to
- Describe the function of the processor.
- Identify the two types of memory.
- Describe the memory designs.
- Convert memory size into actual memory words.
- Identify and explain the two broad categories of memory use.
- Identify different peripheral devices that can be used with a programmable controller.

The processor unit houses the processor/memory module(s), communications circuitry, and power supply. The processor coordinates and controls the operation of the entire programmable controller system, while the power supply furnishes the necessary DC power required by the processor (Figure 3-1).

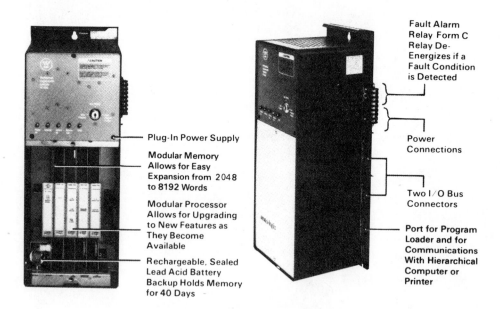

Fault Alarm Relay Form C Relay De-Energizes if a Fault Condition is Detected

Plug-In Power Supply

Modular Memory Allows for Easy Expansion from 2048 to 8192 Words

Modular Processor Allows for Upgrading to New Features as They Become Available

Rechargeable, Sealed Lead Acid Battery Backup Holds Memory for 40 Days

Power Connections

Two I/O Bus Connectors

Port for Program Loader and for Communications With Hierarchical Computer or Printer

Figure 3-1. Westinghouse PC-700 Processor Unit

NOTE: In some cases the power supply will be a separate unit. Depending on the type of memory, volatile or non-volatile, the power supply could also include a backup battery system.

Processors are available that will control as few as 6 and as many as 40,000 "real world" inputs and/or outputs. The size of the processor unit to be used, of course, is dependent upon the size of the process(es) or driven equipment to be controlled. One PC can control more than one machine or process line. The number is limited only by the I/O and memory capacity of the PC used.

NOTE: It is difficult to discuss processor unit configuration due to the differences in PC hardware from the various manufacturers. The discussion that follows is general and is not intended to cover all the PC's on the market today. It should also be noted that when pictures are used to illustrate a given configuration or concept, one of a particular manufacturer's models may be illustrated; however, the manufacturer will also have other models larger and/or smaller in different configurations.

THE PROCESSOR

The processor module which plugs into the processor unit, as stated earlier in Chapter 1, is the "brain" of the programmable controller. The processor consists of the micro-processor, memory chips, circuits necessary to store and retrieve information from the memory, and communication circuits required for the processor to interface with the programmer, printer, and other peripheral devices. The memory and communication circuits can be modules separate from the micro-processor module. The actual hardware configuration will depend on the PC. Figure 3-2 shows a Square D Model 300 processor for smaller systems (128 I/O) where the micro-processor, memory, and communications are all in one module. Figure 3-3 shows the General Electric Model 600 which uses individual modules for such functions as memory, communication, arithmetic.

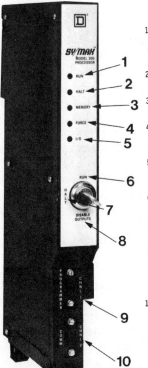

1. RUN (GREEN) - When illuminated, processor is scanning ladder diagram program and operating on I/O's. When flashing, processor is operating on ladder diagram program, but is not energizing any outputs.

2. HALT (RED) - When illuminated or flashing, processor has halted its execution and is no longer scanning program.

3. MEMORY (RED) - When illuminated, the red HALT light will flash, indicating that the processor has halted due to a memory error.

4. FORCE (RED) - When illuminated or flashing, one or several I/O have been forced to an on or off state thereby overriding the ladder diagram program.

5. I/O (RED) - When illuminated, the red HALT light will flash, indicating that the processor has halted due to a malfunction in the I/O system.

6. In this mode, the processor operates on all I/O according to the control program.

7. In this mode, the processor does not operate and all external outputs are turned off.

8. In this mode, the processor operates on the control program; however, all external outputs are disabled. I/O LED's function per the program operation.

9. Communications port to programming equipment.

10. Communications port used for inter-processor communications, access to a computer, or loader/monitor.

Figure 3-2. Square D SY/MAX Model 300 Processor

Figure 3-3. General Electric Model 600 Processor

The **micro-processor** is the device that monitors the state or status (ON-OFF) of the input devices, scans and solves the logic of the user program, and controls the state of the output devices (ON or OFF).

The memory section of the processor consists of the user memory and storage memory. These memories consist of hundreds or thousands of locations where information can be stored. The form in which the information is stored will be covered later in Chapter 5, while the purpose or use of each memory, user and storage, will be covered later in this chapter.

Memory can be separated into two distinct types: **Volatile** and **non-volatile.** A volatile memory is one that loses its stored information when power is removed. Even momentary losses of power will erase any information stored or programmed on a volatile memory chip.

As might be expected, a non-volatile memory has the ability to retain stored information when power is removed, accidentally or intentionally.

To protect a volatile memory, backup batteries are included in the processor power supply. The batteries may be "D" size dry cells, rechargeable nickel cadmium, lead acid or non-rechargeable alkaline and lithium-manganese dioxide types. When batteries are included in the controller power supply, battery indicator lights are also included to indicate the state of the batteries. Common indicator lights would be BAT OK and BAT LOW. Another arrangement would be only one light to indicate that the battery condition is normal. When the light goes out, it is a warning that the batteries need to be replaced. Even after the light goes out, or on other systems when the BAT LOW light comes ON (flashing), the memory will normally still be protected for a minimum of two days. Depending on the size of the memory and the type of batteries used, the memory can remain protected for over one year with fully charged batteries. In reality though, rarely will the power be interrupted and remain OFF for even two days.

The type of batteries used and the number required will vary with each manufacturer. Because the alkaline and lithium batteries are not rechargeable and must be replaced periodically, care must be taken to always replace the batteries with the type specified, paying specific attention to the orientation of each battery in the battery holder to assure that proper polarity is maintained (Figure 3-4).

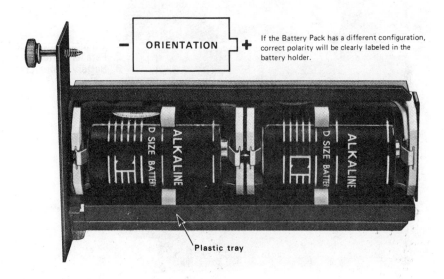

Figure 3-4. Battery Holder
(Courtesy of Allen-Bradley)

CAUTION: Batteries can only be replaced when normal power is being supplied to the controller. Failure to observe this requirement will result in the loss of the user program. As might be expected, there is an exception to the rule. Some memory modules, like the General Electric Series 6 shown in Figure 3-5, use a single lithium manganese dioxide battery that is located on the printed circuit board. Note that a second battery connector has been added and wired in parallel. The new battery is connected to the second connector to protect the memory before the old battery is removed.

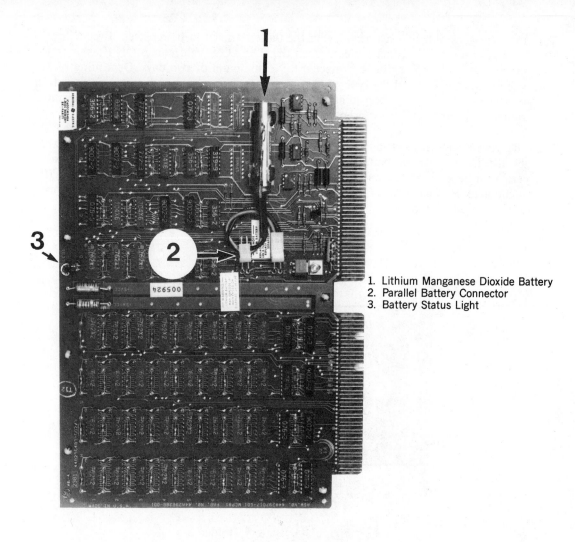

1. Lithium Manganese Dioxide Battery
2. Parallel Battery Connector
3. Battery Status Light

Figure 3-5. Battery Backup for Memory Module
(Courtesy of General Electric)

MEMORY DESIGNS

No attempt will be made to explain solid-state memory design in more than a generalized way for basic understanding. Detailed explanations of solid-state memory design are available in the electronics section of most libraries.

The most common type of volatile memory is **RAM.** Ram stands for read/write solid-state **R**andom **A**ccess **M**emory. Read/write indicates that the information stored in memory can be retrieved or read, while write indicates that the user can program or write information into the memory. The words **random access** refer to the ability of any location (address) in the memory to be accessed or used. Ram memory is used for both the user memory (ladder diagrams) and storage memory in many PC's. Since RAM memory is volatile it must have battery backup to retain or protect the stored program. There are several types of RAM memory: MOS, HMOS, and CMOS to name a few. Of the different types of RAM, the CMOS-RAM (**C**omplimentary **M**etal **O**xide **S**emiconductor) is probably one of the most popular. CMOS-RAM is popular because it has a very low current drain when not being accessed (15 uAmps), and the information stored in memory can be retained by as little as 2V DC. A typical fully charged lithium battery is rated 2.95V at 1.75 amp/hrs.

Non-volatile memories are memories that retain their information or program when power is lost and do not require battery backup. A common type of non-volatile memory is **ROM**. ROM stands for **R**ead **O**nly **M**emory. Read only indicates that the information stored in memory can be read only and cannot be changed. Information in ROM is placed there by the manufacturer for the internal use and operation of the PC, and, of course, the manufacturer does not want the information changed or altered.

Other types of non-volatile memory are PROM, EPROM, EAROM, EEPROM, MAGNETIC CORE, and BUBBLE.

PROM, Programmable **R**ead **O**nly **M**emory, allows initial and/or additional information to be written into the chip. PROM may be written into only once after being received from the PC manufacturer; programming is accomplished by pulses of current. The current melts fusible links in the device, preventing it from being reprogrammed. This type of memory is used to prevent unauthorized program changes.

NOTE: Regardless of the memory type, the memory can also be protected by a key switch located on the front of the processor, or on the programming device. This key switch locks out the programmer. With the programmer "locked-out," the program in the processor can be run but not changed.

EPROM, Erasable **P**rogrammable **R**ead **O**nly **M**emory, is also ideally suited when program storage is to be semi-permanent or additional security is needed to prevent unauthorized program changes. The EPROM chip has a quartz window over a silicon material that contains the electronic integrated circuits. This window normally is covered by an opaque material, but when the opaque material is removed and the circuitry exposed to ultra violet light, the memory content can be erased. The EPROM chip is also referred to as **UVPROM** (Figure 3-6).

Figure 3-6. Typical EPROM Chip
(Courtesy of Allen-Bradley)

The following procedure is used to erase a ladder diagram or other information from an EPROM or UVPROM chip.

1. Carefully and gently rock the EPROM chip out of the ZIF (zero insertion force) socket.
2. Remove the opaque material that covers the quartz window.

CAUTION: Special care and handling of the EPROM, or any IC chip for that matter, must be taken to ensure that the pins do not become dirty or bent.

3. Expose the window to ultraviolet light.

A recommended erasing system for most EPROM chips is the S-11E UV Lamp by Ultra-Violet Products
        1500 Walnut Grove Ave.
        San Gabriel, CA 91778

This system has an intensity of 5000 uW/cm², and will perform a complete memory erasure in approximately 50 minutes. In contrast, if left in bright sunlight, erasure would take about one week. Stronger UV sources would result in shorter exposure times. The user must ensure an adequate exposure dose of at least 15W-SEC/CM². Additional information can be obtained from the PC manufacturer or from the ultra violet lamp supplier.

After erasure, the chip window must once again be covered with an opaque material, such as electrician's tape, to avoid undesirable alteration of the memory.

Once erased, the EPROM chip can be re-programmed using the programmer.

EAROM, **E**lectrically **A**lterable **R**ead **O**nly **M**emory, chips can have the stored program erased electrically. This is accomplished by applying different values of POS(+) and NEG (−) voltage values to specific circuit points. When erasing an EAROM chip, the voltage values and pin location are supplied in the manufacturer's literature. Once erased, the chip can be re-programmed.

EEPROM, **E**lectrically **E**rasable **P**rogrammable **R**ead **O**nly **M**emory, also referred to as E²PROM, is a chip that can be programmed using a standard programming device and can be erased by the proper signal being applied to the erase pin. EEPROM is used primarily as a non-volatile backup for the normal RAM memory. If the program in RAM is lost or erased, a copy of the program stored on an EEPROM chip can be down loaded into RAM.

MAGNETIC CORE Memory uses small magnetic ferrite beads for storing the program or other information. The beads are doughnut shaped and are typically about .050 inches in diameter with about a .030 inch hole. Wires which will carry electrical current are passed through the holes which induce a magnetic field in the ferrite beads or core. The direction of current flow determines the direction of magnetic field. The direction of the magnetic field in each bead can later be read and logic determined.

MAGNETIC BUBBLE Memory (MBM) is one of the newest and most unique types of non-volatile memory. The other memories discussed, except magnetic core, were silicon-based. Bubble memory, however, uses garnet, a mineral often used as the abrasive on sandpaper, and stores information as microscopic magnets. The MBM is constructed of a Gadolinium-Gallium Garnet (GGG) base covered with a thin layer of pure magnetic garnet to form the chip. Strong permanent magnets are placed on top of and underneath the chip with the magnetic flux lines perpendicular to the chip. The strong magnetic flux causes most of the microscopic magnetic domains of the pure garnet to align parallel to and with the same polarity as the flux lines of the permanent magnets. However, many domains in the garnet are left that have an opposite polarity of the magnetic flux. These domains will be shrunk by the external magnet field into tiny round magnetic "Bubbles." By the use of a small electromagnet, bubbles can be removed, moved around in the garnet material, or new bubbles created. For digital storage, a bubble represents one (1), and no bubble in a location represents zero (0).

Bubble memory works in conjunction with standard volatile RAM memory chip(s). When the RAM is initially programmed, the bubble memory acts as a backup. If power is lost or intentionally turned OFF, the RAM loses everything in memory, but the bubble memory retains the original information or program that was in RAM. When power is reapplied, the program is copied from the bubble memory back into RAM.

MEMORY SIZE

PC's are available with memory sizes ranging from as little as 256 words for small systems up to 2M (million) for the larger systems. Memory size is usually expressed in K values: 2K, 4K, 16K, etc. K or Kilo, which usually stands for 1,000, actually represents 1,024 in computerese. The difference between a standard K (1,000) and the 1,024 K value used with processors and computers is due to the binary numbering system that is used. Design engineers wanted to stay with K or Kilo, but as close as they could come with the binary numbering system, Base 2, was 1×2^{10} or 1024. This also explains the reason for the memory size of individual memory chips; 256 (1×2^8), 512 (1×2^9), 1024 (1×2^{10}), etc. A memory of 256 words would be 1/4K, and 64K would actually be 65,536 words. (Words and word structure are covered in Chapter 4.) The actual size of the memory required will, of course, depend on the application. Buying a PC with more memory than is necessary will add significantly to the original cost. In the event that future expansion is planned, there are two options: buy a PC with additional memory greater than current requirements to allow for future expansion, or buy a PC that meets present needs and add memory when the need arises. The memory can be increased by replacing an entire memory module or by adding memory chips to the original memory module. Whichever method is used to increase memory, replacing a module or adding chips to an existing module, will depend upon the PC manufacturer. Figure 3-7 shows a CMOS-RAM memory board. Note the memory chips on the left side of the board.

Figure 3-7. Gould 584 Programmable Controller
with 128K RAM Memory Board

For PC's in which the memory can be expanded by adding volatile CMOS-RAM chip(s) to the memory module, the following procedures should be used:

1. Record a copy of the current user program on disc or magnetic tape depending on the system.
2. Remove main power from the PC.
3. Remove the memory module and take to a clean area.
4. Carefully remove any screws necessary to gain access to the printed circuit board where the extra RAM memory sockets are located. If the backup battery is located on the module, disconnect the battery before removing or installing memory chip(s).

NOTE: The RAM chip(s) will come packaged in a **conductive** plastic bag. Within the bag, each RAM chip will be inserted into a conductive sponge-like material. The conductive, yet highly resistive, material is used to keep all the pins of the chip at the same electrical potential.

While working with the RAM chips, **DO NOT** handle cellophane covered articles such as cigarette packages and candy wrappers, plastic, styrofoam, or other materials that can cause a static charge.

DO NOT install the chip in carpeted areas or in an area with contamination that might foul the pins.

NEVER slide the RAM chip across any surface or store a RAM chip in a non-conductive plastic bag or insert the chip into non-conductive material.

The volatile RAM chips used today are not as susceptible to damage from static charges as they were just a few years ago. But rather than just removing the chip from the conductive material and installing it into the proper socket, the following precautions should still be used if practicable:

1. Use a 3M type 8005* (or equivalent) static-free work station which is connected through a minimum 200K ohm resistor to earth ground.
2. Ground all tools before contacting the RAM chip.
3. Wear a conductive wrist strap which has a minimum 200K ohm resistance and is connected to earth ground.
4. Control relative humidity to 40-60% if possible.

   *Available from 3M Nuclear Products
   3M Center
   St. Paul, MN 55101

Remove the chip from the conductive foam. Be careful to touch only the chip base. **DO NOT** touch the pins.

Inspect the pins for proper alignment. If any pins have been bent, gently straighten them using needle-nosed pliers that have been grounded.

A dot or notch on the case of the chip is used for proper orientation of the chip into the socket.

Grasp the chip by both ends and gently set it in the socket. **DO NOT INSERT.** Be sure the chip is positioned so the dot or notch of the chip matches the dot or notch on the socket.

Before attempting to insert the chip into the socket, check each pin to make sure it lines up properly with the corresponding socket point. Make any necessary pin adjustments as outlined above.

When pin alignment is assured, the chip may be inserted into the socket. Insertion is accomplished by pressing **gently** on the case of the chip until the chip is fully seated into the socket.

Carefully re-assemble the memory module, remembering to reconnect the backup battery if one was mounted on the module, re-install the module in the processor, and re-apply power to the system. The user program can now be re-entered into the processor; plus, any additional user program can now be added using the new memory chip(s).

CATEGORIES OF MEMORY

As indicated earlier, the processor memory can be divided into two broad categories; **user memory** and **storage memory**.

User memory contains the ladder diagram instructions programmed by the user. The instructions can be entered either by a programming device, hand held or desktop type, a magnetic tape, or a system computer. Programming can also be accomplished by using a telephone interface from a remote computer, tape loader, or programming device. The user memory may also store user programmed messages that can be recalled or activated by contacts in the ladder diagram. Upon activation, the message(s) is displayed on the CRT of the programming device or other remote CRT, and can also be printed out on a compatible printer.

Message storage and recall could be used to alert an operator if a bearing is heating up. A typical programmed message could be as follows: Bearing hot on Motor #4. A thermocouple at the bearing of Motor #4 would be tied to one input circuit of a thermocouple input module. When the temperature of the bearing exceeds a pre-determined value, the thermocouple would cause that circuit of the thermocouple input module to close or turn ON. The corresponding contact of the input device that would be programmed in the ladder diagram would close and the message would be generated.

Storage memory is where the status (ON-OFF) of all discrete input and output devices are stored. The numeric values of timer and counters (preset and accumulated), numeric values for arithmetic instructions, and the status of internal relays are also stored in this memory.

As indicated at the start of this chapter, the memory section was presented in a very generalized and broad approach. While the information applies generally to all PC's, more specific information and memory structure can only be obtained by reviewing the specifications and literature of the individual manufacturers. In subsequent chapters the memory structure of specific PC's will be discussed and illustrated, but the text will not cover all the PC's on the market today.

PERIPHERALS

Peripheral is defined as a device connected to a processor to provide an auxiliary or support function.

Such equipment could be a printer (Figure 3-8) for producing hard copy printouts of the user program and/or other processor information.

Figure 3-8. Dot Matrix Printer
(Courtesy of OKIDATA)

Most newer dot matrix printers will require a standard RS-232C 25 pin connector for communicating with the processor. The RS-232C connector may be mounted on the processor unit itself or may be mounted on the back of the programming device (Figure 3-9). In either case, the printer may be activated by a keyed sequence entered from the programming device and/or initiated from the processor itself.

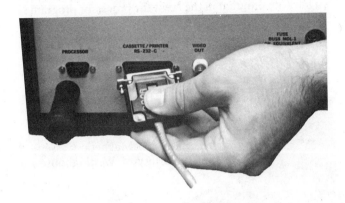

Figure 3-9. RS-232C Connection for Printer/Cassette Recorder
(Courtesy of Square D Company)

Using a compatible computer terminal and the appropriate software enables the user program and data files to be stored on floppy disc and/or hard disc. Figure 3-10 shows an Allen-Bradley T50 industrial terminal. The T50 terminal is an IBM AT compatible unit that will also run other software like Lotus 1-2-3*. Several major PC manufacturers have introduced IBM compatible terminals for use with their standard line of controllers.

*IBM is a registered trademark of the I.B.M. Corporation.
 1-2-3 is a registered trademark of the Lotus Development Corporation.

Figure 3-10. Allen-Bradley T-50 Industrial Terminal with
3½" Floppy Disc Drive and 10 MB Hard Drive

Another peripheral device is the magnetic tape loader. The tape loader is used to record and store the user program or to load pre-programmed instructions into the processor (Figure 3-11). Data quality cassette tapes or data cartridges are used by the tape loader, depending on the type, to record and store the user program. Once the program has been taped, a tab on the plastic body of the cassette or cartridge can be moved, or on some units removed, to prevent recording or "over-writing" on top of an existing recorded program. Figure 3-12 shows both the data cartridge and cassette types of magnetic tape units. Recording the user program provides a back-up program in the event the processor program is lost due to memory failure or accidental erasure.

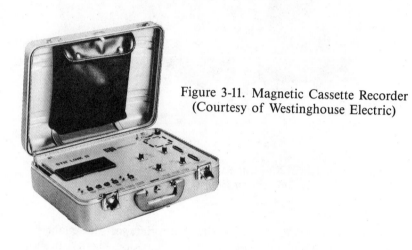

Figure 3-11. Magnetic Cassette Recorder
(Courtesy of Westinghouse Electric)

DATA CARTRIDGE

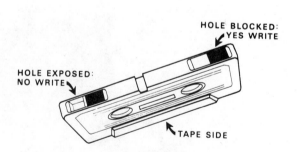

CASSETTE

Figure 3-12. Data Cartridge and Cassette-Type Magnetic Tape Units
(Courtesy of Square D Company and Allen-Bradley)

Tape loaders may be a separate unit as shown in Figure 3-10 or built into a programming device as shown in Figure 3-13.

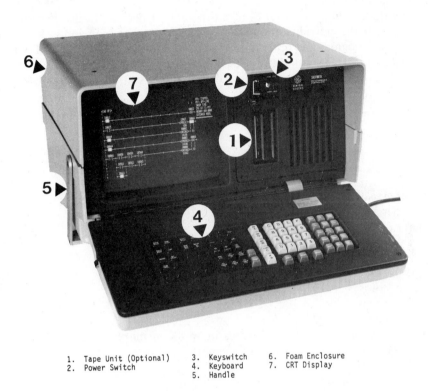

1. Tape Unit (Optional)	3. Keyswitch	6. Foam Enclosure
2. Power Switch	4. Keyboard	7. CRT Display
	5. Handle	

Figure 3-13. Programming Device with Built-In Data Cartridge Loader
(Courtesy of General Electric)

The tape loader, like the printer, may connect by cable directly to an RS-232C connector mounted on the processor or may connect to the back of the programming device (Figure 3-9) depending on the PC.

Some PC's have a communication connector or "port" for direct connection to a computer for reading and/or writing (programming) into the processor memory.

Another peripheral device is the **MODEM** (**MO**dulator and **DEM**odulator). The MODEM can be used to connect the processor via telephone lines to a computer, tape loader, printer, or to another PC.

Chapter Summary

The processor contains the circuitry necessary to monitor the status (ON-OFF) of all inputs and control the condition (ON-OFF) of all the output devices. It also has the ability to solve and execute the individual program steps in the user program and has a memory for storing the user program and other numeric information, as well as the ability to retrieve and use any and all information stored in memory. The processor not only communicates with the I/O rack and the programming device, but it can also communicate with any peripheral device(s) that may be connected.

Review Questions

1. The processor is often referred to as the _____ of the programmable controller.
2. Briefly describe volatile memory.
3. Briefly describe non-volatile memory.
4. 1K of memory is actually:
    a. 1000 words
    b. 1010 words
    c. 1024 words
    d. 1042 words
5. Calculate the actual number of words in an 8K memory.
6. The most common type of volatile memory is:
    a. PROM
    b. EAROM
    c. Bubble
    d. RAM
7. Which of the following are types of non-volatile memory?
    a. EEPROM
    b. PROM
    c. RAM
    d. Bubble
    e. EAROM
    f. Magnetic Core
8. List the two broad categories of memory.
9. Define the word peripheral.
10. List two common peripheral devices.

Chapter 4

Numbering Systems

Objectives

After completing this chapter, you should have the knowledge to
- Understand decimal, binary, binary coded decimal (BCD), hexidecimal, octal, and binary coded (BCO) numbering systems.
- Convert from one numbering system to another.

For the programmable controller to function properly and to control a process or driven equipment, it must be able to perform the user program repeatedly and accurately. The system must also be able to perform its control function with great speed. The required speed is achieved by processing all information in binary signals. A binary signal is represented by two stable states. These states are 1 and 0. The 1 and 0 can represent ON or OFF, voltage or no voltage, high or low, or any other two conditions depending upon the system. The key to the speed with which binary information can be processed is that there are only two states, each of which is distinctly different. There is no in-between state or condition, and, when information is processed, the decision is either YES or NO. There is no MAYBE or any other alternative.

As indicated in Chapter 3, the processor memory consists of hundreds or thousands of locations. These locations are referred to as words. Each word is capable of storing binary data in the form of binary digits or bits. **BITS** is an acronym for **BI**nary digi**TS**. A binary digit, like a binary signal, can only be a 1 or a 0. The number of bits that a word can store will depend on the system or PC. Words can be made up of 32 bits, 16 bits, 12 bits or 8 bits. The 16 bit word is the most common. Figure 4-1 illustrates a 16 bit word.

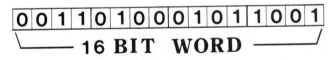

Figure 4-1. 16-Bit Word

If a memory size is 256 words, then it can actually store 4,096 bits of information using 16 bit words (256 words X 16 bits per word) or 2048 bits using an 8 bit word (256 words X 8 bits per word). When comparing memory sizes of different PC systems, it is important to know the number of bits per word of memory.

Bits can also be grouped within a word into **BYTES**. A BYTE is defined as the smallest complete unit of information that can be transmitted to or retrieved from the processor at a given time.

NOTE: The number of BITS that make up a BYTE will depend on the hardware requirements of any given PC system. A BYTE that consists of 8 BITS is very common.

So that information stored in each word can be located, each word is numbered or given an ADDRESS. Addressing words in the memory serves the same function as the addresses used for homes or apartments. As an example, word 100 would represent a specific word location in memory just like N. 100 Lincoln could represent the address for an apartment building. The bits in word 100 could be found by referencing a given bit number, just like an occupant of the apartment complex is found by a given apartment number.

Since a bit of information can only be a 1 or 0 (ON or OFF), how is the status of bits within a word determined? As an example, words that store the status of individual bits for input devices are set to 1 (ON) or 0 (OFF) depending on the status (ON-OFF) of the input devices that the bit locations represent. Other bits are set to 1 or cleared to 0 by the processor in response to the logic of the user program, relay ladder logic, or special instructions, which in turn can control the status (ON or OFF) of other bits that represent output devices.

A simple example of how this works is illustrated in Figure 4-2.

NOTE: The example uses memory organization and addressing used with one of the families of Allen-Bradley PC's. While the example is specific to Allen-Bradley, the concepts illustrated are common to all PC's. Allen-Bradley uses an octal numbering system to address bit locations. Notice that the 16 bits are numbered 00 through 07 and 10 through 17. In the octal system, numbers 8 and 9 are never used. The octal numbering system is covered later in this chapter.

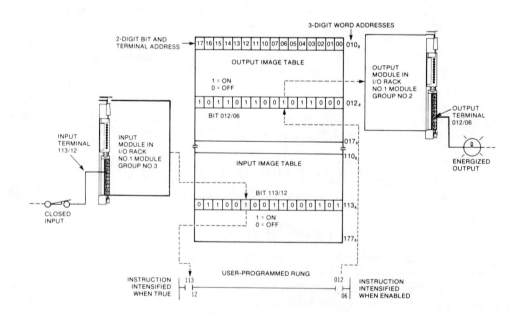

Figure 4-2. Relationship of Bit Addresses to Input and Output Devices
(Courtesy of Allen-Bradley)

44

Assume that when a given limit switch is closed, the closure is to be indicated by having a light come ON. The limit switch is connected to an input module in an I/O rack while the indicator light is connected to an output module. Chapter 2 discussed that the DIP switches were set in a prescribed sequence to identify the I/O rack number for the processor and that the location of each terminal point on each I/O module within the rack determined the address of a given device. In Figure 4-2 the limit switch is connected to terminal 11312 on an input module or has an address of 11312. This actually indicates that bit 12 of word 113 will store the status (ON-1 or OFF-0) of the limit switch. The indicator lamp is connected to terminal 01206. This address (01206) indicates that bit 06 of word 012 will control the status (1-ON or 0-OFF) of the lamp.

By programming a simple circuit into the user memory of the processor as shown at the bottom of Figure 4-2, the processor controls the indicator lamp using the logic of the program. The logic says if contact 11312 CLOSES, lamp 01206 should light or go ON. When power is applied to the processor, the processor looks at bit 12 of word 113 to see if the bit is set to 1 or to 0. If the limit switch is open, the bit will be set to 0 or OFF. If the limit switch is closed, as indicated in Figure 4-2, the input module sends a signal to the processor, and bit 12 of word 113 will be set to ON or 1. As the logic of the ladder diagram indicates that when contact 11312 (bit 12 of word 113) is CLOSED or ON that the indicator lamp 01206 must also be ON, the processor reads the logic and sets bit 06 of word 012 to 1, which, in turn, turns ON the lamp connected to the output module.

The address 11312 also tells us that the limit switch is an input device and is wired to terminal 12 of module group 3 in rack 1.

Allen-Bradley uses 5 digit addresses on many of their PC's. Figure 4-3 illustrates the significance of each digit or group of digits.

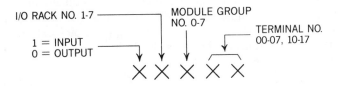

Figure 4-3. Allen-Bradley 5-Digit Address Format

The first digit is used to indicate whether the address is for an input or output. The number 1 represents an input while an 0 represents an output.

The next, or second digit, identifies the rack number. This will always be a number from 1–7 (octal numbering). The next number identifies the module group within the rack. This will always be a number from 0–7. The last two digits identify the actual terminal number that the device is wired to.

Figure 4-4 reviews the concept using the address of limit switch 11312.

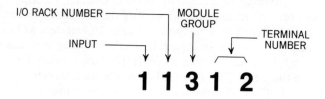

Figure 4-4. Limit Switch Address 11312

A 1 in the first digit location tells us that the address represents an input device. The next digit, also a 1, tells us that the device is located in I/O rack number 1. The next digit, which is a 3, further identifies the location as module group number 3. The last two digits, 1 and 2, identify the actual terminal (12) that the limit switch is connected to.

A review of the concept is shown in Figure 4-5. The limit switch address 11312 gives us a hardware location for an input device in rack 1, module group 3, terminal 12. This same address 11312 tells us that the status or ON-OFF state of the limit switch will be reflected by bit 12 of word 113.

This same addressing scheme will give us a hardware location for the indicator lamp addressed 01206.

The 0 in the first digit location indicates an output device. The next digit being a 1 tells us that the I/O rack location is rack 1. The next digit identifies the module group as group 2. The last two digits locate terminal 06 as the terminal that the indicator lamp is wired to. Again, the address 01206 also locates the memory word and bit location that will reflect the status (ON-OFF) of the indicator lamp as shown in Figure 4-6.

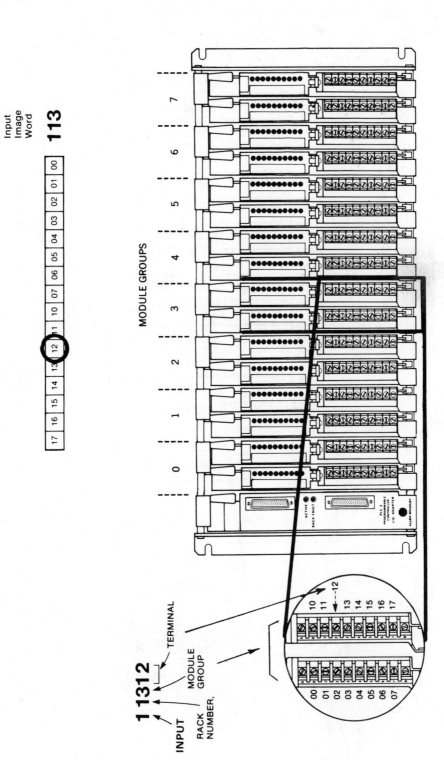

Figure 4-5. Relating Input Address 11312 to Actual Hardware Location
(Courtesy of Allen-Bradley)

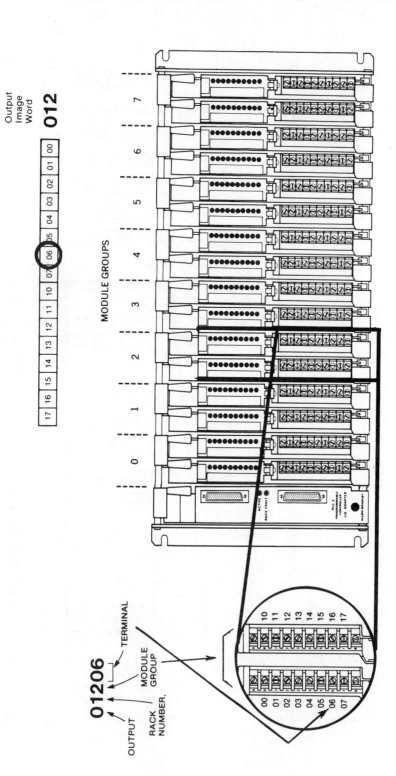

Figure 4-6. Relating Output Address 01206 to Actual Hardware Location
(Courtesy of Allen-Bradley)

48

For larger systems that use more than 7 I/O racks, a 7 digit number is used as shown in Figure 4-7.

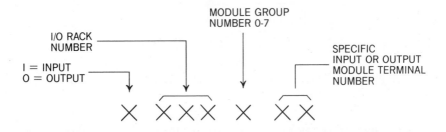

Figure 4-7. Allen-Bradley 7-Digit Address Format

While the address system discussed is specific to Allen-Bradley, most PC manufacturers use an addressing scheme that not only identifies memory word locations, but will also give hardware location.

Storing and retrieving information using numerical information or digits is called a **digital** system.

Information other than the status of inputs and outputs can also be stored using only 1's and 0's. Examples of other information include pre-set and accumulated values for timers and counters and arithmetic functions. But before further discussing how information is stored and retrieved, the numbering systems used by the different PC manufacturers need to be understood.

PC manufacturers use several numbering systems to convert decimal numeric information into binary digits for memory storage and control of outputs. These numbering systems include the following: binary, binary coded decimal (BCD), hexadecimal, octal, and octal coded decimal (OCD). To better understand these numbering systems, the most common and familiar system, the decimal numbering system, will be reviewed.

DECIMAL SYSTEM

In the decimal system, ten unique numbers or digits 0 through 9 are used. When a numbering system uses ten digits, it is called base 10. The value of a decimal number depends on the digit(s) used and the place value of each digit. In the decimal system, the first position to the left of the decimal point is called the units place, and any digit from 0 to 9 can be used (Figure 4-8). The next position to the left of units is the tens place, then hundreds, thousands, etc., with each place extending the capability of the decimal system by ten times or a power of ten. Each place position can also be represented as a power of ten, starting with 10^0 (Figure 4-8).

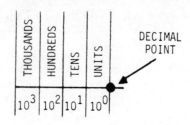

Figure 4-8. Place Value and Corresponding Power of Ten

A specific decimal number can be expressed by adding the place values as shown in Figure 4-9.

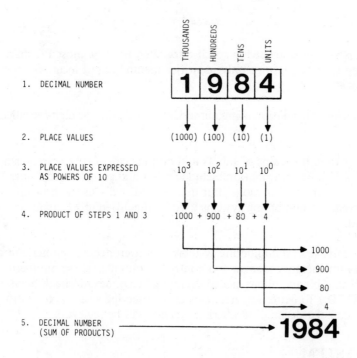

Figure 4-9. Decimal Numbering System

Mathematically, each place value is expressed as a digit number times a power of the base or 10, for the decimal numbering system.

Another example using the decimal number 239 is shown in Figure 4-10.

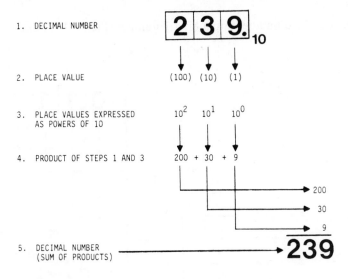

Figure 4-10. Decimal Numbering System

BINARY SYSTEM

The binary or base 2 system uses only two digits: the numbers 1 and 0. This system is used to store information in the processor memory in the form of BITS (binary digits). Like the decimal system, or all numbering systems for that matter, each digit has a certain place value. The first place to the left of the starting point (binary point), like in the decimal system, is the units or 1's location (base 2^0). The next place to the left of the units place is 2's or base 2^1 as shown in Figure 4-11. The next place value is 4's or base 2^2, then 8's or base 2^3, etc. A binary number will always be indicated by placing a 2 to the right of the units digit. Figure 4-11 illustrates how a binary number is converted to a decimal equivalent number.

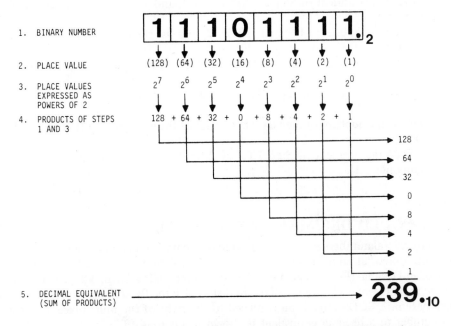

Figure 4-11. Converting a Binary Number to a Decimal Number

51

For converting a decimal number into a binary number (Figure 4-12), the following procedure is used.

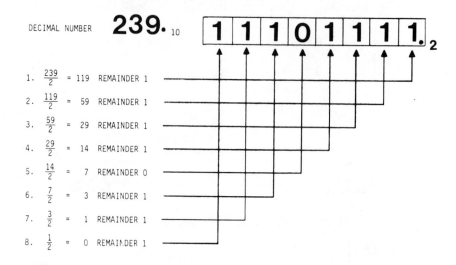

Figure 4-12. Converting a Decimal Number to a Binary Number

Step 1: The decimal number is divided by 2 (base of binary numbering system). The quotient is listed (119) as well as the remainder (1).

Step 2: Divide the quotient of Step 1 (119) by base 2, and list the new quotient (59) and the remainder (1).

Step 3: Divide the quotient of Step 2 (59) by base 2, and list the new quotient (29) and remainder (1).

Step 4: Divide the quotient of Step 3 (29) by 2, and list the new quotient (14) and the remainder (1).

Step 5: Divide the quotient of Step 4 (14) by 2, and list the new quotient (7) and remainder (0).

Step 6: Divide the quotient of Step 5 (7) by 2, and list the new quotient (3) and remainder (1).

Step 7: Divide the quotient of Step 6 (3) by 2, and list the new quotient (1) and remainder (1).

Step 8: Divide the quotient of Step 7 (1) by 2, and list the new quotient (0) and the remainder (1).

BINARY CODED DECIMAL (BCD) SYSTEM

When large decimal numbers are to be converted to binary for memory storage, the process can become somewhat cumbersome. To solve this problem and to speed conversion, the Binary Coded Decimal (BCD) system was devised. In the BCD system, 4 binary digits (base 2) are used to represent each decimal digit. To distinguish the BCD numbering system from a binary system, a BCD designation is placed to the right of the units place. Converting a BCD number to a decimal equivalent is shown in Figure 4-13.

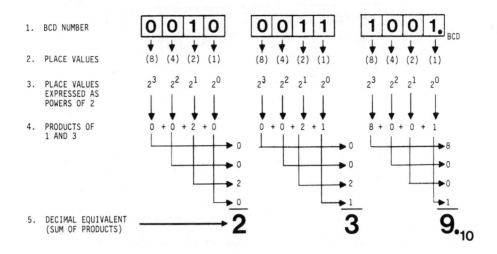

Figure 4-13. Converting a BCD Number to a Decimal Number

When using a BCD numbering system, 3 decimal numbers may be displayed using 12 bits (3 groups of 4), or 16 bits (4 groups of 4) may be used to represent 4 decimal numbers or digits.

When only 3 decimal digits are to be represented, using 12 bits, they may be further identified as Most Significant Digit (MSD), Middle Digit (MD) and Least Significant Digit (LSD) as shown in Figure 4-14.

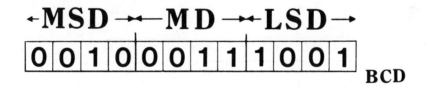

Figure 4-14. Identifying BCD Digits

By using the BCD system, the largest decimal number that can be displayed by any four binary digits is 9. Figure 4-15 shows the four binary digit equivalents for each decimal number 0 through 9.

PLACE VALUE				DECIMAL EQUIVALENT
2^3 (8)	2^2 (4)	2^1 (2)	2^0 (1)	
0	0	0	0	0
0	0	0	1	1
0	0	1	0	2
0	0	1	1	3
0	1	0	0	4
0	1	0	1	5
0	1	1	0	6
0	1	1	1	7
1	0	0	0	8
1	0	0	1	9

Figure 4-15. Binary to Decimal Equivalents
(Courtesy of Allen-Bradley)

HEXADECIMAL SYSTEM

The hexadecimal system, often referred to as just HEX, consists of a number system with base 16. The HEX system is used when large numbers need to be processed. The hexadecimal system is also used by some PC's for entering output instructions into a sequencer.

It would seem logical that the numbers used in base 16 would be 0 through 15. But in reality only numbers 0 through 9 are used, and the letters A-F are used to represent numbers 10-15 respectively. The place values from the hexadecimal point are 1's-16^0, 16's-16^1, 256's-16^2, 4096's-16^3, etc.

Each hexadecimal digit is represented by four (4) binary digits. The binary equivalents are shown in Figure 4-16.

HEXADECIMAL	BINARY	DECIMAL
0	0000	0
1	0001	1
2	0010	2
3	0011	3
4	0100	4
5	0101	5
6	0110	6
7	0111	7
8	1000	8
9	1001	9
A	1010	10
B	1011	11
C	1100	12
D	1101	13
E	1110	14
F	1111	15

Figure 4-16. Hexadecimal Equivalents for Binary and Decimal
(Courtesy of Allen-Bradley)

54

The decimal number 4780 is converted to hexadecimal as illustrated in Figure 4-17.

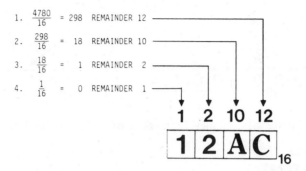

Figure 4-17. Converting a Decimal Number to a Hexadecimal Number

Step 1:   The decimal number is divided by 16 (base for hexadecimal numbering system). The quotient is listed (298) as well as the remainder (12). On a calculator the answer is 298.75. The quotient then is 298, and the remainder is .75 X 16 or 12.

Step 2:   Divide the quotient of Step 1 (298) by 16, and list the new quotient (18) and the remainder (10). On a calculator the answer is 18.625. The quotient then is 18, and remainder is .625 X 16 or 10.

Step 3:   Divide the quotient from Step 2 (18) by 16, and list the new quotient (1) and the remainder (2). On a calculator 18 divided by 16 equals 1.125. The quotient then is 1, and the remainder is .125 X 16 or 2.

Step 4:   Divide the quotient from Step 3 (1) by 16, and list the new quotient (0) and the remainder (1). On a calculator 1 divided by 16 equals .0625, so the quotient is 0, and the remainder is .0625 X 16 or 1.

Converting a hexadecimal number to a decimal number is illustrated in Figure 4-18.

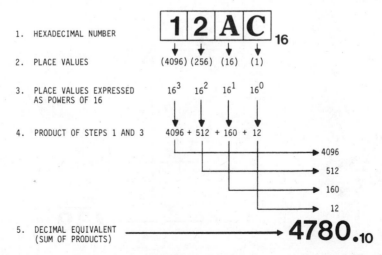

Figure 4-18. Converting a Hexadecimal Number to a Decimal Number

NOTE: Remember that A is equivalent to 10 and C is equivalent to 12.

The binary equivalent of the hexadecimal number 12AC is shown in Figure 4-19.

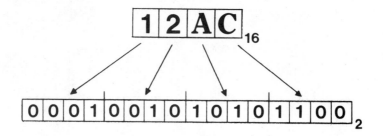

Figure 4-19. Binary Equivalent of a Hexadecimal Number

OCTAL SYSTEM

The octal system or base 8 is made up of eight digits: numbers 0 through 7. The first digit to the left of the octal point is the units or 1 and has a base or power of 8^0. The next place is eights (8's) or base 8^1. The next place would be sixty-fours (64's) or base 8^2, followed by five hundred twelves (512's) or base 8^3, 4096's or base 8^4, etc. An octal number will always be expressed by placing an eight to the right of the units digit as shown in Figure 4-20.

$$357._8$$

Figure 4-20. Octal Number

The process for converting an octal number to a decimal equivalent number is illustrated in Figure 4-21.

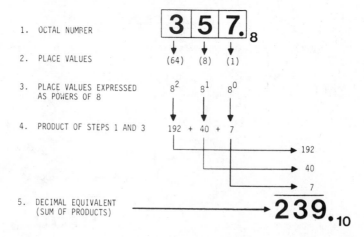

Figure 4-21. Converting an Octal Number to a Decimal Number

Because the largest single number that can be expressed using the octal numbering system is seven (7), each octal digit can be represented by using only three (3) binary bits.

Converting a number from the octal numbering system to the binary system is illustrated in Figure 4-22.

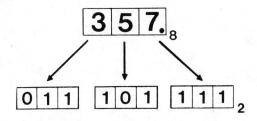

Figure 4-22. Conversion of Octal Number to Binary

Allen-Bradley, for example, uses the octal numbering system for address words and bit locations in their data table or memory. When using the octal numbering system, words are labeled 000 to 007, 010 to 017, 020 to 027, etc., and the terminals on their output modules are labeled 00-07 rather than 1-8.

BINARY CODED OCTAL (BCO) SYSTEM

The Binary Coded Octal (BCO) system uses eight bits (binary digits) to represent 3-digit octal (base 8) numbers from $000._8$ to $377._8$. The eight bits are broken down into three groups: 2 bits, 3 bits and 3 bits as shown in Figure 4-23.

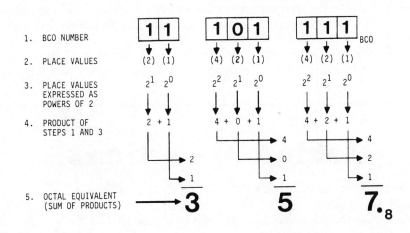

Figure 4-23. BCO Numbering System

57

With this arrangement of binary digits, the largest number that could be obtained if all bits were set to 1 is 377 (Figure 4-24).

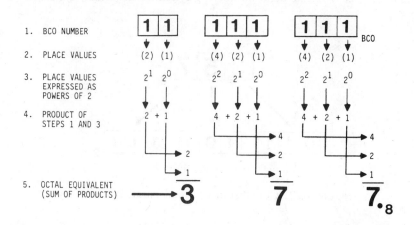

Figure 4-24. Converting BCO to the Equivalent Octal Number

BCO numbers are converted to decimal equivalent as illustrated for the octal numbering system in Figure 4-21.

Chapter Summary

There are several numbering systems that can be used to store information in the form of binary digits (BITS) into the memory system of a processor. Which specific numbering system is used or the combination of numbering systems will depend on the hardware requirements of the specific PC manufacturer. The important thing to remember, however, is that no matter which numbering system or systems are used, the information is still stored as 1's or 0's.

Review Questions

1. When information is stored using only 1's and 0's it is called a _____ system.
2. A BIT is an acronym (initials) for _____.
3. The decimal numbering system uses 10 digits or a base 10. List the base for each of the following numbering systems.
   a. Binary   Base _____
   b. Hexadecimal   Base _____
   c. Octal   Base _____
4. Convert binary number 11011011 to a decimal number.
5. Convert decimal number 359 to binary.

6. Convert binary coded decimal (BCD) number 0101 0011 1001 to a decimal number.
7. Convert hexadecimal number 14CD to a decimal number.
8. Convert decimal number 3247 to a hexadecimal number.
9. Convert decimal number 232 to octal.
10. How do we prevent binary numbers 10 or 11 from being confused as decimal numbers 10 and 11?
11. Convert the following binary values to decimal.
    10011000
    01100101
    10011001
    00010101
12. Convert the following BCD values to decimal.
    1001 1000
    0110 0101
    1001 1001
    0001 0101
13. The BCD values 1001 0011 0101 is NOT:
    a. 935 decimal
    b. 0011 1010 0111 Binary
    c. 647 octal
    d. 3A7 hexadecimal
14. The hexadecimal value 2CB is NOT:
    a. 715 decimal
    b. 1313 octal
    c. 0010 1100 1011 binary
    d. 0111 0001 0011 BCD

Chapter 5

Memory Organization

Objectives

After completing this chapter, you should have the knowledge to
- Identify the two broad categories of memory and describe the function of each.
- Identify the types of information stored in each category of memory.

Chapter 4 covered the various numbering systems used by the different PC manufacturers for entering data into memory. Chapter 5 can now look at actual memory organization.

At this point it is no surprise to find out that not all PC manufacturers have organized their memories in the same way or that they do not all use the same terminology for the configuration or make-up of their memories.

As discussed in Chapter 3, there are two broad categories of memory: user memory and storage memory (Figure 5-1).

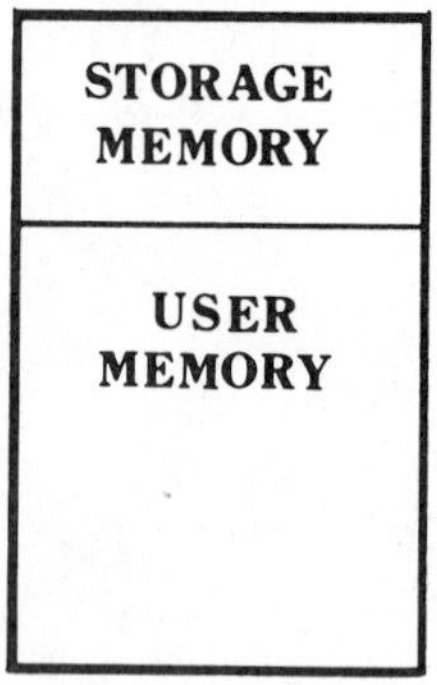

STORAGE MEMORY
USER MEMORY

Figure 5-1.

STORAGE MEMORY

Storage memory is that portion of memory that will store information on the status of discrete input and output devices, preset and accumulated values of timers and counters, internal I/O relay equivalents, numerical values for arithmetic functions, and so on. The entire storage memory may be called a data table, a register table, or another name depending on the PC manufacturer. A register is defined as an area for storing information, logic or numeric. Although the names or titles which are given to sections or subsections of the storage memory vary, the principles involved do not.

As an example, the section of the memory that stores the status of the discrete or "real world" input devices may be referred to as an input image table, input register, input status table, or external input section. No matter what name is used, the information is stored in the same way. The status (ON or OFF) of each input device is stored as either a 1 or 0 (ON or OFF) in one bit of either an eight or 16 bit word. When the processor is executing the user program (ladder diagram) it scans the input device status stored in the storage memory to determine which inputs are ON or OFF.

The section of storage memory set aside for discrete output status may be referred to as the output image table, output register, output status table, or external output section. Again, the name does not change the function of this section of the storage memory or the method that information is placed in memory for control of the actual output devices. As the processor executes the user program, it sends binary data (1's or 0's) to the output section of memory to control the output devices. Each output device is represented by one bit of a memory word.

Numeric information for timer/counter preset and accumulated values, arithmetic functions, sequencer functions, data manipulation, and so forth uses a part of the storage memory that may be called data registers or internal storage. Information is entered and stored in this part of memory using the binary, BCD, hexadecimal, octal, or BCO numbering systems. The numbering system(s) used, will, of course, depend on the PC hardware and system requirements. As indicated earlier, numeric information requires that several bits be used of one word to represent numbers. In a practical sense, then, any word used to store numerical information is not available for additional storage, even if all the bits of a word are not used.

Any complete words that are not used for storing numeric values can be used as internal relays. Internal relays will replace the numerous control relays used in most "hard wired" control circuits. Internal or dummy relays may also be programmed in the I/O section of memory when all the words or bits of words are not being used for discrete I/O devices. Some PC's have a portion of memory set aside just for internal relays. The concept and use of internal relays is covered later in the text.

Figure 5-2 shows the address table for the storage memory section of a Square D Company Model 300 processor.

EXTERNAL I/O	01-01	01-02	01-03	01-04	01-05	01-06	01-07	01-08	01-09	01-10	01-11	01-12	01-13	01-14	01-15	01-16
	02-01	02-02	02-03	02-04	02-05	02-06	02-07	02-08	02-09	02-10	02-11	02-12	02-13	02-14	02-15	02-16
	03-01	03-02	03-03	03-04	03-05	03-06	03-07	03-08	03-09	03-10	03-11	03-12	03-13	03-14	03-15	03-16
	04-01	04-02	04-03	04-04	04-05	04-06	04-07	04-08	04-09	04-10	04-11	04-12	04-13	04-14	04-15	04-16
	05-01	05-02	05-03	05-04	05-05	05-06	05-07	05-08	05-09	05-10	05-11	05-12	05-13	05-14	05-15	05-16
	06-01	06-02	06-03	06-04	06-05	06-06	06-07	06-08	06-09	06-10	06-11	06-12	06-13	06-14	06-15	06-16
	07-01	07-02	07-03	07-04	07-05	07-06	07-07	07-08	07-09	07-10	07-11	07-12	07-13	07-14	07-15	07-16
	08-01	08-02	08-03	08-04	08-05	08-06	08-07	08-08	08-09	08-10	08-11	08-12	08-13	08-14	08-15	08-16
INTERNAL I/O (Relay Equivalents)	09-01	09-02	09-03	09-04	09-05	09-06	09-07	09-08	09-09	09-10	09-11	09-12	09-13	09-14	09-15	09-16
	10-01	10-02	10-03	10-04	10-05	10-06	10-07	10-08	10-09	10-10	10-11	10-12	10-13	10-14	10-15	10-16
	11-01	11-02	11-03	11-04	11-05	11-06	11-07	11-08	11-09	11-10	11-11	11-12	11-13	11-14	11-15	11-16
	12-01	12-02	12-03	12-04	12-05	12-06	12-07	12-08	12-09	12-10	12-11	12-12	12-13	12-14	12-15	12-16
	13-01	13-02	13-03	13-04	13-05	13-06	13-07	13-08	13-09	13-10	13-11	13-12	13-13	13-14	13-15	13-16
	14-01	14-02	14-03	14-04	14-05	14-06	14-07	14-08	14-09	14-10	14-11	14-12	14-13	14-14	14-15	14-16
	15-01	15-02	15-03	15-04	15-05	15-06	15-07	15-08	15-09	15-10	15-11	15-12	15-13	15-14	15-15	15-16
	16-01	16-02	16-03	16-04	16-05	16-06	16-07	16-08	16-09	16-10	16-11	16-12	16-13	16-14	16-15	16-16
DATA REGISTERS	17-															
	18-															
	19-															
	110-															
	111-															
	112-															

Figure 5-2. Square D Model 300 Storage Memory Addresses

The discrete inputs and outputs would be addressed 01-01 (word one-bit one) through 08-16 (word 8-bit 16) for a total of 128 I/O. Words 9 (bits 1-16) to word 16 (bits 1-16) are for internal I/O (relay equivalents), and words 17-112 are for numerical data storage. As mentioned earlier, any unused external I/O words can be used for internal relays or data storage. Any unused internal I/O words can be used for data storage, and any unused data storage words can be used for internal I/O (internal relays). While unused external I/O words can be used for storage, unused storage words cannot be used for external or discrete I/O.

The complete memory for the Model 300 also has user memory and additional memory for processor use.

Figure 5-3 shows the memory organization for an Allen-Bradley Mini PLC-2/15. Allen-Bradley uses a 16 bit word. The word and bit addresses are numbered using the octal numbering system as shown in Figure 5-3. The first eight words (000-007) of the data table are set aside for the processor. Words 010-017 and 020-026 are the output image table. An output device addressed 01000 would be bit 00 of word 010; likewise, an address of 02617 would be bit 17 of word 026. Note that part of word 027 is used by the processor for battery low condition, message generation, and data highway. The next 40 word section of memory, words 030-037, 040-047, 050-057, 060-067, and 070-077, is used for storing timer/counter accumulated values (numeric) or for internal storage.

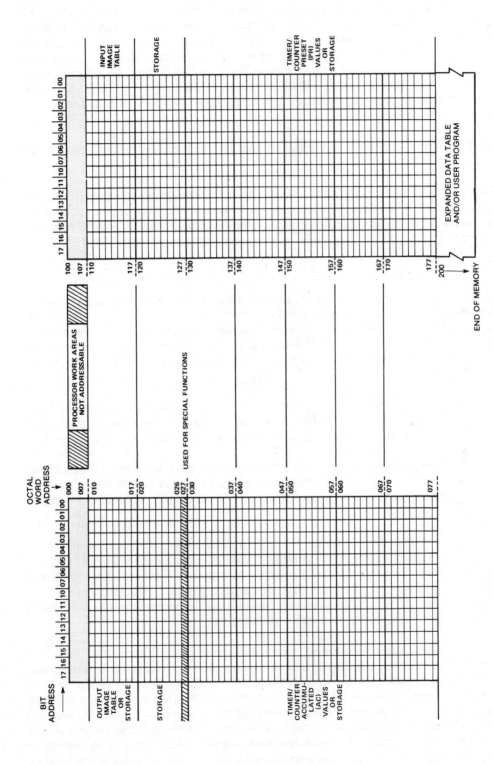

Figure 5-3. First 128 Words of an Allen-Bradley PLC 2/15 Memory

The next 8 words are the second processor work area. Words 110 through 127 (16 words) are the input image table for discrete input addresses. Input address 12406 would be bit 06 of word 124. Words 125 and 126 are used by the processor to indicate remote I/O faults. Words 130-177 (40 words) are used to store the preset values of timers/counters or for internal storage. If word 030 is used for a timer, word 130 automatically stores the preset value of timer 030. The accumulated value for timer 030 is stored in word 030. Any timer or counter in an Allen-Bradley system will store the preset value in the word that is 100 higher than the timer or counter number. For example, counter 047 would have its preset value stored in word 147.

Below the timer/counter preset values and internal storage section is the user program section of memory.

The actual configuration of the data table can be changed to meet user needs. Input/output image tables can be increased or expanded to handle more discrete inputs and outputs. Additional information on expanding the data table of Allen-Bradley PC's is available from their local representative or from their technical publications.

USER MEMORY

The user memory, or logic memory as it is sometimes called, is where the programmed ladder diagram is entered and stored. Within the user memory are words set aside as **holding** registers. Holding registers typically store information generated and used by the processor when it is solving the user program. Holding registers that are set aside to store intermediate values or other short-term bits of information are sometimes referred to as scratch areas or scratch pads.

The user memory may account for most of the total memory of a given PC system. A system with an 8K memory (8192 words) typically would have a storage memory of 2K or less, and the balance of memory (6K) would be available for user memory.

Once the user program has been entered into the user memory, by either a programming device, tape loader, computer, or via telephone interface, the programmable controller would be ready to control the process or driven equipment in accordance with the user program logic.

When the PC is powered-up or turned ON, the first thing the processor typically does is run an internal self diagnostic or self check. Typically, if any part of the processor system is not functioning, the processor fault light or other indicator light will come ON. With some systems, if the programming device is connected, a written explanation or fault code will appear on the CRT screen. Some systems even indicate the part number of the integrated circuit or printed circuit board that has malfunctioned.

When the processor has passed the self diagnostic check, it is ready to go to work. The first thing a typical processor does is look at or scan the input image table to determine the status of all discrete input devices, checking for 1's and 0's. Next the processor scans the user memory and solves the logic of the user program. Once the user program is solved, the processor scans the output image table and sets the appropriate bits to ON, which in turn, through the output modules in the I/O rack, turn on the required discrete output devices. The processor then immediately goes back and looks at the input image table again to determine the status of the discrete input devices. Remember that the status is determined by a 1 or a 0 in the word and bit location that corresponds to the input device address, depending on whether the device is ON or OFF. The processor then scans the user memory again, solving the program logic, and then scans and sets the bits in the output image table, if necessary, to turn ON or OFF discrete output devices.

The time it takes the processor to complete one scan (read inputs, solve logic, enable or disable outputs) is VERY fast, in the milli (m) seconds. Since one scan happens so rapidly, scanning can also be thought of as I/O update-solve logic, I/O update-solve logic, and so on. The actual scan time required, of course, will vary with the number of I/O devices and size of user program.

A rule of thumb for scan time is 1 to 8m seconds per K words of user program. As with most rules of thumb, this is a very general estimate and is given only to indicate how rapidly the processor can scan the program and update the I/O section. The scanning rate is so fast that, in most instances, it will appear to the operator that all logic is solved simultaneously, when, in fact, the logic is scanned sequentially. Scanning starts with the first element of the first network in user memory and solves each subsequent network in order. Special programming techniques and functions can alter the scanning order, but for standard circuits the scan will be in network order. The specific scan times on any given PC can be obtained from the PC manual or technical literature.

Chapter Summary

The processor memory (storage and user) stores the I/O status, the user program, and numeric data used by the processor. All data, logic and numeric, is stored with binary digits which are represented as either a "1" or a "0." By storing binary data, the processor can rapidly scan and execute the user program and update the I/O section. The names given to memory sections or subsections may be unique to each PC manufacturer, but the memories all work in basically the same manner.

Review Questions

1. The following types of information are normally found and/or stored in one of the PC's two memory categories (storage and user). Place an S (storage memory) or a U (user memory) before the information type to indicate in which memory it is normally found and/or stored.
   a. _____ Status of discrete input devices
   b. _____ Preset values of timers and counters
   c. _____ Ladder diagram
   d. _____ Numeric values for arithmetic
   e. _____ Holding registers
2. T   F   When a PC is first turned ON it will run a self diagnostic or self check test.
3. Describe the three steps of a typical PC scan.
4. T   F   The actual scan time, or time it takes the PC to complete a 3 step scan will decrease as the number of program words increases.

Chapter 6
Understanding and Using Ladder Diagrams

Objectives

After completing this chapter, you should have the knowledge to
- Identify a wiring diagram.
- Identify the parts of a ladder diagram.
- Convert a wiring diagram to a ladder diagram.
- List the rules that govern a ladder diagram.

There are basically two types of electrical diagrams; wiring diagrams and ladder diagrams.

WIRING DIAGRAMS

The **wiring** diagram shows the circuit wiring and its associated devices (relays, timers, motor starters, switches, etc.) in their relative physical location (Figure 6-1). While this type of diagram assists in locating components and to show how a circuit is actually wired, it does not show the circuit in its simplest form. To simplify understanding of how a circuit works and to show the electrical relationship of the components (not the physical relationship) a **ladder** diagram is used.

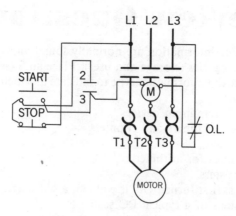

Figure 6-1. Wiring Diagram

LADDER DIAGRAMS

The ladder diagram, also referred to as a schematic diagram or elementary diagram, is used by the technician to speed his understanding of how a circuit works. Figure 6-2 shows the same circuit as Figure 6-1 but in ladder diagram form.

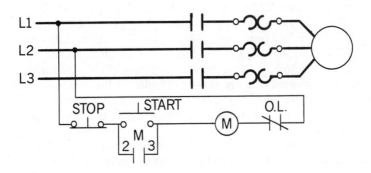

Figure 6-2. Ladder Diagram

To simplify the circuit and to speed understanding, the power portion of the circuit is shown separate from the control portion and no attempt is made to show actual physical location of the components. Because the motor connections or power portion would be the same for any three phase motor, it is common practice to not even show the motor starter or the motor. By not showing the power portion of the circuit, a ladder diagram showing only the control portion of the diagram would appear as shown in Figure 6-3.

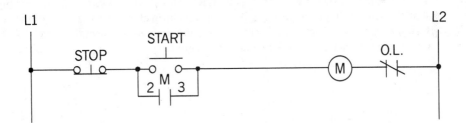

Figure 6-3. Simplified Ladder Diagram

The power required for the control circuit is always shown as two vertical lines, while the actual line(s) of logic are drawn as horizontal lines. The power lines, or rails as they are often called, are like the vertical sides or rails of a ladder, while the horizontal logic lines are like the rungs of a ladder.

Refer back to Figure 6-1. While it is easy to determine the physical relationship between the stop-start station, the motor starter coil (M), the overload contacts (O.L.) and the holding contacts (2 and 3), it is difficult to determine the electrical relationship. The ladder diagram in Figure 6-3, however, clearly shows the electrical relationship between all of the control circuit components.

LADDER DIAGRAM RULES

Some basic rules for ladder diagrams are as follows:

1. A ladder diagram is read like a book, left to right and top to bottom.
2. The vertical power lines or rails of the ladder diagram represent the voltage potential of the circuit. The potential could be AC or DC and vary in voltage from 6 to 480V. Standard labeling for the rails is L1 and L2. L1 is AC high or hot for AC circuits and positive or plus (+) for DC circuits. L2 is AC low or neutral for AC circuits and negative or minus (−) for DC circuits. The rails may also be marked X1 and X2 when the voltage potential is derived from a transformer.
3. Devices or components are shown in order of importance whenever possible. Notice in Figure 6-3 that the stop button is shown ahead of the start button. For safety reasons the stop button has a higher order of importance than the start button.
4. Electrical devices or components are shown in their normal condition. Normal for electrical diagrams is with the circuit de-energized or OFF. The stop button is shown closed since that is normal for a stop button. The holding contacts (2 and 3) of coil M are shown open. Again, this is the normal position for these contacts when coil M is de-energized. The normally open (N.O.) M holding contacts 2 and 3 will not close until there is a complete path for current flow to coil M. When coil M energizes, M contacts 2 and 3 will close providing a parallel path for current flow with the start button.
5. Contacts associated with relays, timers, motor starters, etc. will always have the same number or letter designation as the device that controls them. This labeling method holds true no matter where in the circuit the contact(s) appear. For example, in Figure 6-3, the N.O. holding contacts 2 and 3 are controlled (activated) by motor starter coil M, therefore, the contacts are identified with a letter M.
6. All contacts associated with a device will change position when the device is energized. Figure 6-4 shows a control relay (CR) controlled by a switch (S-1) on rung 1 of the ladder diagram. Rung 2 shows a normally closed (N.C.) CR contact in series with a green indicator lamp, while rung 3 shows a normally open (N.O.) CR contact in series with a red indicator light.

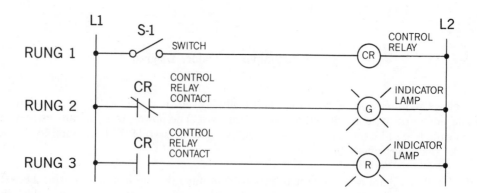

Figure 6-4. Three Rung Ladder Diagram

When power is applied to the rails of the ladder diagram, the only device in the circuit that will operate is the green indicator lamp. The green lamp lights due to a complete path for current flow through the N.C. (normally closed) CR (control relay) contacts. These contacts are normally closed and will only change position and open when the control relay in rung 1 is energized. When switch S-1 is closed, completing the path for current flow and energizing the CR in rung 1, the N.C. CR contacts in rung 2 will open, while the N.O. CR contacts in rung 3 go closed. The action of the contacts will turn OFF the green lamp in rung 2 and turn ON the red lamp in rung 3. As long as the control relay remains energized through S-1 the normally closed contact in rung 2 will remain open and the normally open contact in rung 3 will remain closed. When S-1 is opened and CR de-energizes the contacts controlled by CR will return to their normal state (N.C. in rung 2 and N.O. in rung 3).

7. In a ladder diagram, devices that perform a stop function are normally wired in series. Figure 6-5 shows two switches wired N.C. that control a green indicator lamp.

Figure 6-5. Two Switches Wired in Series

With the two switches wired in series, both A and B must remain closed for the lamp to remain lit. If either switch is opened, the green lamp will go out. When switches and/or contacts are wired in series they are said to have an **AND** relationship. The AND relationship requires that both A and B must be closed for the lamp to light. A truth table for this concept is shown in Figure 6-6.

SWITCH	SWITCH	INDICATOR LAMP
A	B	G
OFF	OFF	OFF
OFF	ON	OFF
ON	OFF	OFF
ON	ON	ON

Figure 6-6. Truth Table for Series Devices

8. Devices that perform a start function are normally wired in parallel. Figure 6-7 shows two switches (A and B) wired in parallel to control a red indicator lamp. In this configuration, if either switch A or B is closed the red lamp will light.

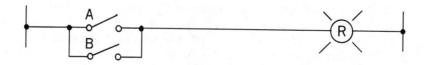

Figure 6-7. Two Switches Wired in Parallel

When switches or contacts are wired in parallel they are said to have an **OR** relationship. The OR relationship requires that either A or B (or both) be closed for the red indicator lamp to light. A truth table for this concept is shown in Figure 6-8.

SWITCH	SWITCH	INDICATOR LAMP
A	B	R
OFF	OFF	OFF
OFF	ON	ON
ON	OFF	ON
ON	ON	ON

Figure 6-8. Truth Table for Parallel Devices

With a better understanding now of what a ladder diagram is and the rules that apply, a discussion of the basic motor stop-start circuit shown in Figure 6-2 will add to your understanding.

BASIC STOP/START CIRCUIT

As stated earlier in this chapter, the wiring diagram in Figure 6-1 is great for showing actual physical location of the circuit wiring and components but it does not show the electrical relationship of the devices as simply as does the ladder diagram. While the wiring diagram will be used for original installation and for some troubleshooting, the ladder diagram will be used to show the electrical relationship of the components and to speed understanding of how the circuit works.

By viewing the ladder diagram in Figure 6-9, it can be seen that when power is applied to the circuit, the motor starter coil M cannot energize because there is an incomplete path for current flow due to the open start button and also the N.O. M contacts (2 and 3). The start button and the N.O. M contacts are wired in parallel and have an OR relationship. When the start button is pushed, a path for current exists from L1 potential through the N.C. stop button, through the now closed start button, through the coil of the motor starter (M), through the N.C. overload contacts to L2 potential.

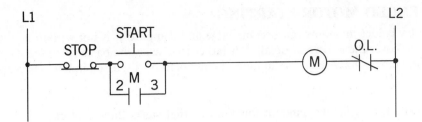

Figure 6-9. Ladder Diagram for Basic Stop/Start Circuit

When the starter coil M energizes, the M contacts (2 and 3) will close providing an alternate path for current flow. At this point the start button could be released and the circuit would remain energized, or held in, by the holding contacts (2 and 3) of the motor starter. When contacts from a motor starter or other device are wired in this fashion they are often referred to as holding, maintain, or sealing contacts as the circuit is held, maintained, or sealed in after the start button is released.

When the holding contacts (2 and 3) are closed, the main motor contacts of the motor starter are also closed and the motor is started. Again, the operation of the motor is normally taken for granted and will not be shown on the ladder diagram. By keeping the ladder diagram as simple and uncluttered as possible, the relationship between components and how the control portion of the circuit works is greatly enhanced.

Figure 6-10 again shows the wiring diagram of a motor stop-start circuit. While this diagram looks entirely different from the ladder diagram, they are electrically the same. This comparison should show you why the ladder diagram is the preferred diagram for electricians and/or technicans.

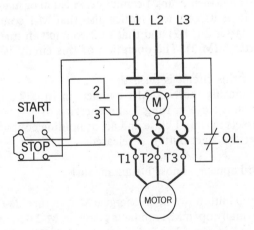

Figure 6-10. Wiring Diagram for Basic Stop/Start Circuit

The ladder diagram has been the working language of electricians and electrical engineers for many years and helps explain why most PC's are programmed using ladder logic. While this method of programming was welcomed by some, it has frustrated other PC users who have not been exposed to or trained in relay ladder logic.

SEQUENCED MOTOR STARTING

While relay ladder diagrams can become large and complex, it is not within the scope of this text to cover them in great detail, but rather to discuss the basic rules and to present some concepts to enhance understanding of circuits to be discussed in later chapters.

Figure 6-11 shows a ladder diagram for a circuit that starts three motors.

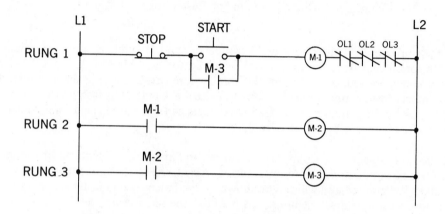

Figure 6-11. Three Motor Start Circuit

Rung 1 contains the stop-start buttons and the motor starter coil M-1 for motor 1. Notice that the holding contacts wired in parallel with the start button are not M-1 contacts, but M-3 contacts. With this arrangement, rung 1 cannot be sealed in or maintained unless motor starter 3 energizes and closes its contacts. Notice also that M-1 contacts in rung 2 must close to energize motor starter 2 (M-2) and that M-2 contacts in turn must close in rung 3 to energize motor starter 3 (M-3). The operation of this circuit is as follows.

When the start button of this circuit is pushed:

1. M-1 energizes closing the N.O. M-1 contacts in rung 2 and energizes M-2.
2. M-2 N.O. contacts close in rung 3 and energize M-3.
3. M-3 N.O. contacts in rung 1 close and act as holding contacts to keep the circuit energized after the start button is released.

NOTE: This sequence happens almost instantaneously.

4. Pushing the stop button would de-energize M-1, in turn de-energize M-2 in rung 2 when the normally open M-1 contacts go open. M-2 de-energizing would open the M-2 contacts in rung 3 and de-energize M-3. With M-3 de-energized, the N.O. M-3 contacts in rung 1 would open. By wiring all three overload contacts in series with M-1 in rung 1, an overload on any motor would shut down all motors. An open overload contact would have the same effect as pushing the stop button.

It could be said that this circuit consists of basically three elements.

1. Inputs
2. Outputs
3. Logic

The inputs are the stop button, start button, and overload contacts. The outputs are motor starters M-1, M-2, and M-3. The logic that caused the sequential starting were normally open contacts M-1, M-2, and M-3.

These three elements — inputs, outputs, and logic — also work well with PC's. The inputs will be wired to input modules, the outputs are wired to output modules, and the processor will perform the logic functions.

Figure 6-12 is the wiring diagram for the 3 motor circuit just discussed. This diagram should further illustrate the point that while wiring diagrams are great for giving you physical location, they do not show the control function of the circuit as clearly as a ladder diagram does.

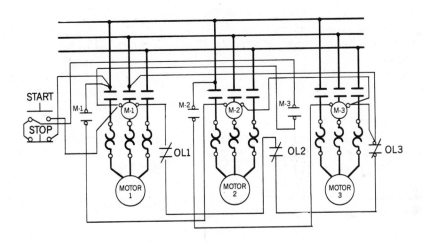

Figure 6-12. Wiring Diagram for Three Motor Circuit

Chapter Summary

There are basically two types of electrical diagrams: wiring diagrams and ladder diagrams. Wiring diagrams show actual physical location and wiring while ladder diagrams show electrical relationship. The simplified ladder diagram speeds understanding of circuit operation and is used for circuit design and troubleshooting. The vertical sides of the ladder diagram are referred to as rails, while the horizontal lines or logic are called rungs. On electrical diagrams, devices are always shown in their normal or de-energized condition. When two or more devices are wired in series they perform an AND function, while two or more devices wired in parallel perform an OR function. The elements of the ladder diagram are inputs, outputs, and logic.

Review Questions

1. Define normally open and normally closed.
2. Describe the difference between a wiring diagram and a ladder or schematic diagram.
3. Explain the operation of circuit 6-9 if M contacts 2 and 3 do not close.
4. Contacts wired in parallel have what relationship?
   a. AND
   b. OR
5. Contacts wired in series have what relationship?
   a. AND
   b. OR
6. The two main vertical lines of a ladder diagram are often referred to as:
   a. Rungs
   b. Power ports
   c. Rails
   d. Tracks
   e. None of the above
7. The horizontal lines or logic of a ladder diagram are referred to as:
   a. Rungs
   b. Power ports
   c. Rails
   d. Tracks
   e. None of the above
8. Devices that are intended to perform a stop function are normally wired in_____
   _____ with each other.
9. Devices that are intended to perform a start function are normally wired in_____
   _____ with each other.
10. How are contacts that are associated with relays, motor starters, timers, etc., identified?
11. Convert wiring diagram 6-A to a ladder diagram.

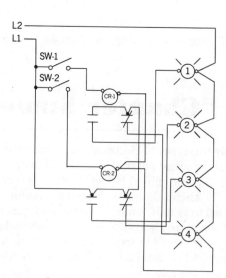

Figure 6-A

12. Convert wiring diagram 6-B to a ladder diagram.

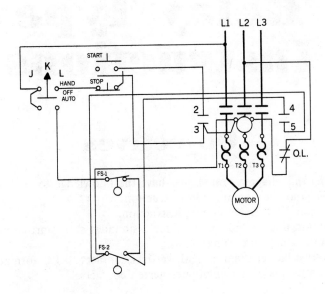

Figure 6-B

Chapter 7

Relay Type Instructions

Objectives

After completing this chapter, you should have the knowledge to
- Understand the Examine On instruction.
- Understand the Examine Off instruction.
- Write and understand the logic for a standard stop/start circuit.
- Describe a network (rung).
- Convert a ladder diagram with vertical contacts to PC format.
- Write a program that eliminates nested contacts.

With most of the background information covered, the actual programming (entering the ladder diagram) of a circuit into the processor so the PC can control the circuit(s) can be discussed. The actual programming will be accomplished using either a desktop or hand held programming device. Figures 7-1a and b show a Modicon P370 hand held programmer for programming and monitoring the Modicon Micro 84 PC.

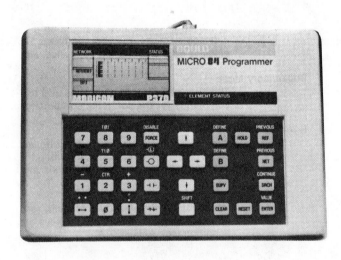

Figure 7-1a. Modicon P370 Hand Held Programmer
(Courtesy of Gould, Inc.)

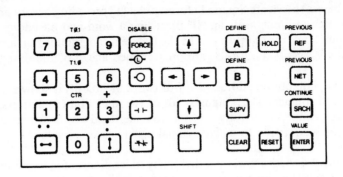

Figure 7-1b. Modicon P370 Keyboard
(Courtesy of Gould, Inc.)

Regardless of the type of programmer used, some common relay symbols are standard. These symbols include normally open contacts ┤├, normally closed contacts ┤╱├, and coil or output ─○─ . Notice in Figure 7-1b that the keyboard has no symbols for input devices such as stop buttons, limit switches, and pressure switches. Contacts from input devices will be programmed using either the N.O. or N.C. relay contact symbols. Actual programming devices by some of the different PC manufacturers will be covered in Chapter 8, but first the relay logic used for PC's must be discussed and **UNDERSTOOD.**

PROGRAMMING CONTACTS

Your first thought is probably: Isn't relay logic—relay logic? The answer is both yes and no. To understand this apparent contradiction, refer to the ladder diagram for a standard stop/start station, Figure 7-2a and the equivalent diagram programmed for a PC, Figure 7-2b.

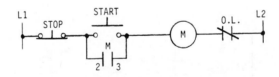

Figure 7-2a. Standard Stop/Start Ladder Diagram

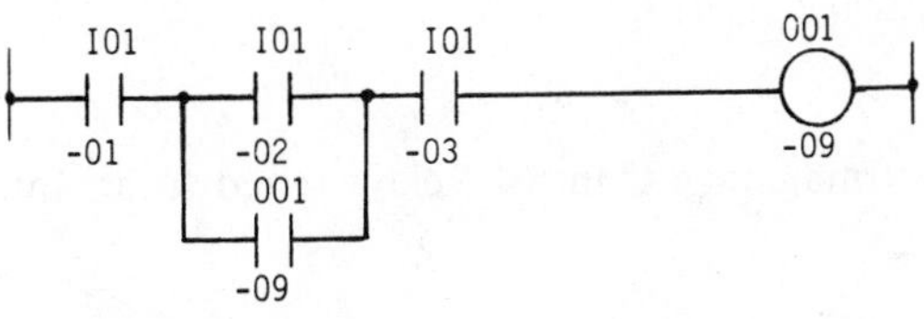

Figure 7-2b. Equivalent Stop/Start Circuit Programmed
with a PC

NOTE: Addresses shown are Square D format. An "I" preceding a word and bit number indicates an input while an "0" preceding a word and bit address indicates an output. Since the output must be the last item programmed on a rung, the overload contacts (O.L./I01-03) are programmed ahead of the output.

In Figure 7-2b the stop button addressed I01-01 (word 01 - bit 01) is shown normally open, as are the overload contacts (O.L.), address I01-03 (word 01 - bit 03). Although this circuit appears to be programmed incorrectly, it is not. To understand why the circuit is programmed this way, a technique referred to as "relay analogy" is used.

Remember, all input devices are wired individually to terminals of an input module (Figure 7-3) and are not actually wired in series as shown in Figure 7-2a.

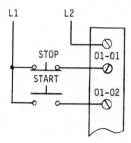

Figure 7-3. Discrete Inputs wired to an Input Module

For the sake of discussion, imagine that each input device is connected to an invisible control relay inside the input module and that each control relay has one normally open and one normally closed contact as shown in Figure 7-4.

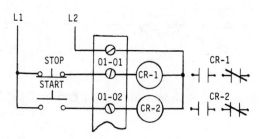

Figure 7-4. Imaginary Control Relays wired to an Input Module

78

Next, connect one lamp to the normally open contacts and another lamp to the normally closed contacts of CR-1 as shown in Figure 7-5.

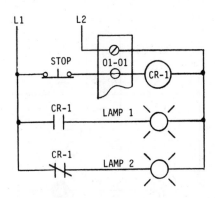

Figure 7-5. Lamps wired to N.O. and N.C. CR-1 Contacts

When power is applied to L1 and L2, CR-1 energizes through the normally closed contacts of the stop button. With CR-1 energized, normally open contacts of CR-1 close, and lamp 1 will light as indicated in Figure 7-6. The normally closed contacts of CR-1 are now open, so lamp 2 cannot light.

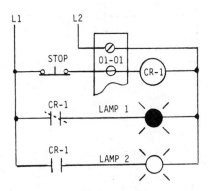

Figure 7-6. Lamp 1 lights with power applied to L1 and L2

Using the invisible relay analogy, normally open CR-1 contacts controlled by a normally closed pushbutton will CLOSE or CONDUCT when power is applied to the circuit. Likewise, normally open (N.O.) contacts **programmed** to represent a normally closed (N.C.) pushbutton will conduct when power is applied to the PC.

A normally open start button connected to an input terminal and an invisible or imaginary control relay as shown in Figure 7-7 would not light lamp 1 until the start button was depressed and would only stay lit as long as the button was held down. Lamp 2 would light as soon as power was applied, but would go out when the start button is depressed and CR-2 energizes (Figure 7-8).

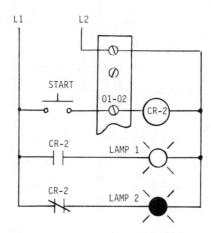

Figure 7-7. Power applied—Start Button not Depressed

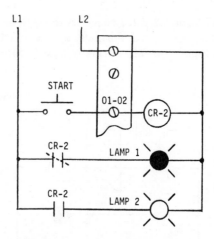

Figure 7-8. Power applied—Start Button Depressed

The rules for contacts that represent discrete or "real world" input devices are shown in Figures 7-9a and b.

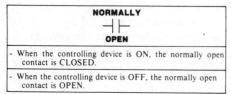

Figure 7-9a.
(Courtesy of Square D Company)

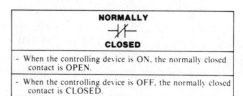

Figure 7-9b.
(Courtesy of Square D Company)

Of course, there are no invisible control relays in the input modules, but there are also no symbols on the programming device for stop buttons, start buttons, limit switches, and the like. As long as relay contact symbols must be used in place of regular input symbols, the relay analogy is then an easy way to explain why normally OPEN (N.O.) contacts are programmed to represent normally CLOSED (N.C.) input devices.

Another approach for understanding how contacts must be programmed is to look at the actual BITS that represent the input devices. Once the input devices are wired to the input module(s) and the PC system is "powered up" or turned ON, the processor scans the inputs and sets the corresponding BITS to 1 or 0 depending on the state of the discrete input device. If an input is OPEN the corresponding bit is set to 0 or OFF, while any bit that represents a CLOSED device will be set to 1 or ON.

In Figure 7-10 the stop button (I01-01) and the O.L. contact (I01-03) are again programmed as normally opened contacts, but as the stop button and overload contacts are actually closed, bits 01 and 03 of word 01 are set to 1 or ON. Figure 7-11a shows the bit status of word 01 with the processor in the RUN mode.

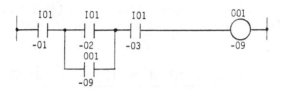

Figure 7-10. Basic Stop/Start Circuit

16	15	14	13	12	11	10	9	8	7	6	5	4	3	2	1
0	0	0	0	0	0	0	0	0	0	0	0	0	1	0	1

Figure 7-11a. Bit Status of Word 01 with processor
in RUN Mode

Even though N.O. contacts were programmed for the stop button and the O.L. contacts, bits 01 and 03 are set to 1 or ON and we need only press the start button to complete the circuit.

When the start button (I01-02) is depressed, the processor scans the I/O section and updates the I/O register setting bit 02 to 1 or ON (Figure 7-11b).

16	15	14	13	12	11	10	9	8	7	6	5	4	3	2	1
0	0	0	0	0	0	0	0	0	0	0	0	0	1	1	1

Figure 7-11b. Bit Status of Word 01 while Start Button
is being Depressed

With the circuit logic now complete, or TRUE, the processor sets bit 09 of word 01 to 1 or ON (Figure 7-11c), and output 001-09 is ENERGIZED and held energized by holding contacts 001-09. Note that the holding contacts 001-09 and the output 001-09 have the same address. They not only have the same address, but they are also the same bit (bit 09 of word 01) in the I/O register. The holding contacts do not actually exist, but the logic equivalent of holding contacts (bit 09 of word 01) set to 1 or ON indicates that output 001-09 is also ON.

16	15	14	13	12	11	10	9	8	7	6	5	4	3	2	1
0	0	0	0	0	0	0	1	0	0	0	0	0	1	0	1

Figure 7-11c. Bit Status of Word 01 after Output 001-09
is Energized and Start Button is Released

When the stop button (input I01-01) is depressed, the processor clears bit 01 to 0, and the circuit logic is broken or goes FALSE. Bit 09 is cleared to 0, and the discrete output device connected to terminal 01-09 would drop out or turn OFF. Word 01 in the I/O register would now appear as shown in Figure 7-11d, with only bit 03, the overload contacts, set to 1.

16	15	14	13	12	11	10	9	8	7	6	5	4	3	2	1
0	0	0	0	0	0	0	0	0	0	0	0	0	1	0	0

Figure 7-11d. Bit Status of Word 01 with Stop Button Depressed

When the stop button is released or CLOSED again, bit 01 would again be set to 1 or ON (Figure 7-11e). The output (001-09) would not energize, however, as the start button (bit 02) is 0 or OFF, and the holding contacts (bit 09) have also been cleared to 0 or OFF.

16	15	14	13	12	11	10	9	8	7	6	5	4	3	2	1
0	0	0	0	0	0	0	0	0	0	0	0	0	1	0	1

Figure 7-11e. Bit Status of Word 01 with Stop Button Released

A third, and last, method of understanding why a normally OPEN symbol is used to represent a normally CLOSED input device is to think of the N.O. symbol ⊣ ⊢ as an EXAMINE ON instruction for the processor. (Figure 7-12).

EXAMINE ON

Figure 7-12. Examine ON Instruction

Whenever the processor sees a N.O. contact in the user program, it views the contact symbol as a request to **EXAMINE** the address of the contact for an **ON** condition. If the N.O. contact has an input address and if the discrete input device is closed or ON, the procesor would set the appropriate bit in the input register to 1 or ON. As the examine ON instruction is looking for an ON condition, a bit set to 1 or ON is a true condition and power can flow through the contacts. If the discrete input had been open or OFF, the processor would have cleared the appropriate bit to 0 or OFF, and the contact would be FALSE as far as the logic of the ladder diagram is concerned and would not allow power flow. By using the control relay analogy again (Figure 7-13a and b), an EXAMINE ON instruction can be better understood.

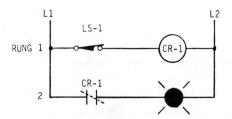

Figure 7-13a. Examine On Instruction with Discrete Input
Device Closed or ON

With the discrete input (LS-1) closed or ON, CR-1 would energize and CR-1 N.O. contacts in rung 2 would close. Rung 2 would now be TRUE and the lamp would light.

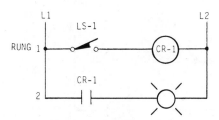

Figure 7-13b. Examine ON Instruction with Discrete Input Device Opened or OFF

If the discrete input (LS-1) is open or OFF, CR-1 is de-energized and the N.O. contacts in rung 2 are open. Rung 2 would be FALSE and the lamp could not light.

The N.C. relay symbol is used as an EXAMINE OFF instruction (Figure 7-14).

EXAMINE OFF

Figure 7-14. Examine OFF Instruction

When the N.C. symbol is programmed in a ladder diagram, the processor views it as a request to **EXAMINE** the address of the contact for an **OFF** condition. Any address that is actually OFF becomes logically TRUE and power can flow. If a N.C. contact has an input address and the discrete input device is OPEN, or OFF, the processor would have set the bit to 0 or OFF. As the EXAMINE OFF instruction is looking for an OFF condition, a bit set to 0 or OFF is a true condition and power can flow through the contact. If the discrete device had been CLOSED, or ON, the bit would have been set to 1 or ON. The EXAMINE OFF instruction can only be logically true when an OFF condition exists. Any bit set to 1 is viewed as an ON condition which makes an EXAMINE OFF (N.C.) contact FALSE and no power can flow. Again, the relay analogy helps to clarify an EXAMINING OFF instruction (Figures 7-15a and b).

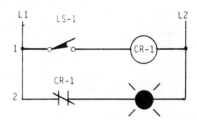

Figure 7-15a. Examine OFF Instruction with Discrete
Input Device Open or OFF

When the discrete input (LS-1) is open or OFF, CR-1 cannot energize. Rung 2 N.C. CR-1 contacts remain closed and the lamp is turned ON.

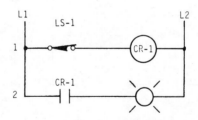

Figure 7-15b. Examine OFF Instruction with Discrete
Input Device Closed or ON

If the discrete input (LS-1) is closed or ON, CR-1 energizes and the N.C. contacts in rung 2 open. Rung 2 is now false and the lamp is turned off.

It would appear that all discrete input devices, whether they are normally open (N.O.) or normally closed (N.C.), are programmed as N.O. contacts to achieve the desired results in the ladder diagram. For many circuit applications this is true, but a big advantage of this programming technique is the ability to turn single pole input devices into double, three or four pole devices in the circuit. Figure 7-16 shows a double pole pressure switch controlling two outputs, motor 1 and motor 2.

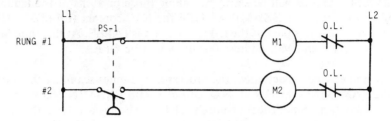

Figure 7-16. Double Circuit Pressure Switch

When power is applied to the circuit, Motor 1 will start through the N.C. contacts of PS-1 in rung 1. Motor 2 cannot start however, due to the N.O. contacts of PS-1 in rung 2. When PS-1 is actuated, the N.C. contacts in rung 1 will open and M1 will go OFF while the N.O. contacts in rung 2 will close, turning M2 ON.

By programming the same circuit on a PC, the necessity of buying a double circuit pressure switch is eliminated. One (1) N.O. and 1 N.C. contact having the same address will be used as shown in Figure 7-17a. The address will actually be the address of a discrete N.C. single circuit pressure switch as illustrated in Figure 7-17b.

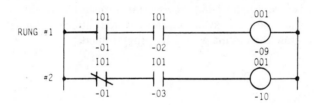

Figure 7-17a. Double Circuit Pressure Switch Circuit Programmed for a Typical PC

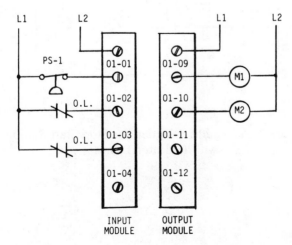

Figure 7-17b. Actual Wiring of a Single Pole Pressure Switch

85

When the processor is placed in the RUN mode, it will EXAMINE all N.O. contacts for an ON condition or EXAMINE ON. As PS-1 and both overload contacts are CLOSED or ON, bits 01, 02, and 03 will be set to 1, making those portions of the ladder diagram TRUE. When the processor EXAMINES OFF the N.C. contact (I01-01), it will see that bit 01 is set to 1 so that this part of the ladder diagram is FALSE. At this time M1, address 001-09, would be ON and M2, address 001-10, would be OFF.

When the pressure switch is actuated, the processor continues scanning the user program and is still EXAMINING for ON all N.O. contacts, and EXAMINING for OFF all programmed N.C. contacts. Since the pressure switch (PS-1) is now actuated, the N.C. contacts are now open, so the N.O. contacts (I01-01) go FALSE turning OFF M1 (001-09), while the N.C. contacts (I01-01) now go TRUE, and turn ON M2 (001-10).

Even though only 2, 3, and 4 poles were mentioned, there is no limit (except user memory size) to the number of times an input device can be addressed and used in a programmed circuit. This programming technique would allow for 6 pole, 7 pole, 8 pole, etc., devices to be programmed using only a single pole discrete device.

There are many more applications and circuits that normally required 2 or 3 pole devices which now will only require single pole devices when programmed for a PC, such as double pole limit switches for forward and reversing circuits and double pole pressure switches for duplex controllers.

When programming contacts (N.O. or N.C.) which are controlled by outputs, the familiar standard relay logic is used. Figure 7-18a shows a standard stop/start station with pilot lights. Lamp 1 (green) indicates power is available, lamp 2 (red) indicates the circuit is activated. Figure 7-18b shows how the circuit would be programmed.

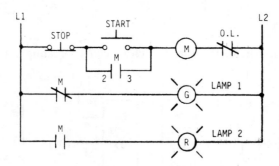

Figure 7-18a. Ladder Diagram for Stop/Start Station
with Indicator Lamps

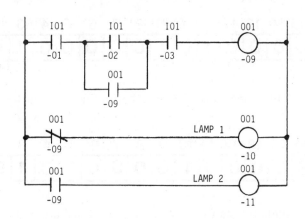

Figure 7-18b. Programmed Circuit for Stop/Start
Station with Indicator Lamps

When the processor is placed in the run mode, lamp 1 will light due to the N.C. contacts 001-09. When the start button is depressed, output 001-09 energizes, N.C. contacts 001-09 OPEN, or go false, and lamp 1 (001-10) goes out. N.O. contacts 001-09 CLOSE, or go TRUE, and lamp 2 (001-11) turns ON, and the holding contacts (001-09) also go true to complete the circuit logic.

Notice that the output, holding contacts, N.C. contact, and N.O. contact for the lamps all have the same address (001-09). The address refers to bit 09 of word 01, and this same bit (09) is referenced four times in the ladder diagram. The EXAMINE OFF or N.C. contact is logically true when bit 09 is cleared to 0 or OFF, while the EXAMINE ON or N.O. contacts will not be logically true until bit 09 is set to 1 or ON.

Figure 7-19a through e show the bit status for the circuit with power applied (Figure 7-19a), start button depressed (Figure 7-19b), start button released (Figure 7-19c), stop button depressed (Figure 7-19d), and after stop button is released (Figure 7-19e).

16	15	14	13	12	11	10	9	8	7	6	5	4	3	2	1
0	0	0	0	0	0	1	0	0	0	0	0	0	1	0	1

Figure 7-19a. Bit Status of Word 01 with Power Applied

16	15	14	13	12	11	10	9	8	7	6	5	4	3	2	1
0	0	0	0	0	1	0	1	0	0	0	0	1	1	1

Figure 7-19b. Bit Status of Word 01 with Start Button Depressed

16	15	14	13	12	11	10	9	8	7	6	5	4	3	2	1
0	0	0	0	0	1	0	1	0	0	0	0	0	1	0	1

Figure 7-19c. Bit Status of Word 01 with Start Button Released

16	15	14	13	12	11	10	9	8	7	6	5	4	3	2	1
0	0	0	0	0	0	1	0	0	0	0	0	0	1	0	0

Figure 7-19d. Bit Status of Word 01 with Stop Button Depressed

16	15	14	13	12	11	10	9	8	7	6	5	4	3	2	1
0	0	0	0	0	0	1	0	0	0	0	0	0	1	0	1

Figure 7-19e. Bit Status of Word 01 with Stop Button Released

LIMITATIONS OF LADDER DIAGRAM

Each PC manufacturer has a limit to the number of contacts or other logic symbols that can be included on one line of a ladder diagram and also a limit to the number of parallel branches or lines that make up one **network**. A network is defined as a group of connected logic elements used to perform a specific function. Figure 7-20 shows a typical network consisting of series contacts and three parallel branches. A network also constitutes one rung of a ladder diagram.

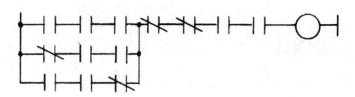

Figure 7-20. Network (Rung)

A typical network limitation of 10 series contacts per line and 7 parallel lines or branches is typical of Westinghouse and Square D Company PC's as shown in Figure 7-21. Also, there is a further limitation with some PC's of only one output per rung or network, and the output must be on the first line.

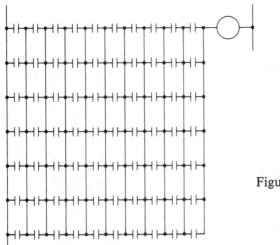

Figure 7-21. Network Limits

NOTE: Special function and other logic symbols will alter the network limitations and requirements; check the operation or program manual for additional information.

General Electric has a 9 X 8 **Matrix** (network) limitation (Figure 7-22), while Allen-Bradley has an 11 X 7. Both manufacturers also limit outputs to one per network, and that output must be on the first line of the rung.

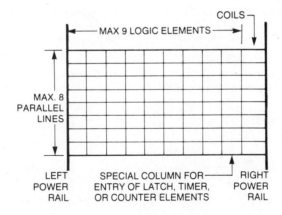

Figure 7-22. General Electric Matrix (Network) Limitation

NOTE: While the number of elements and lines within a network is limited, the only limit on the number of networks or rungs is the size of user memory.

Other PC's, like Modicon 484, have networks that allow for more than one output (parallel outputs), as shown in Figure 7-23.

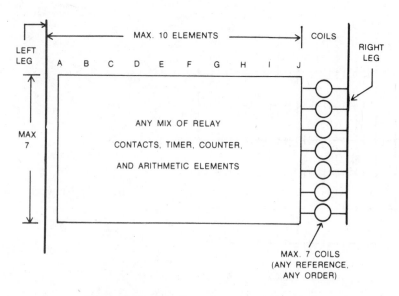

Figure 7-23. Modicon Program Format

When a circuit requires more series contacts than the network allows (Figure 7-24a), the contacts are split into two rungs (Figure 7-24b). The first rung would contain part of the required contacts and would be programmed to an internal or "dummy" relay. Internal relays are actually a bit and word location in storage memory or an unused bit in the I/O table.

The address of the internal relay, 02000 in the example, would also be the address of the first N.O. contact on the second rung. The remaining contacts (8-13) would be programmed, followed by the address of the discrete output device. When the first 7 contacts close, internal output, bit 00 of word 20, would be set to 1. This would make the N.O. contacts 02000 in rung 2 TRUE, and if the other 6 contacts (8-13) were closed, the rung would be TRUE, and the discrete output would be turned ON.

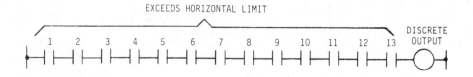

Figure 7-24a. Contacts Exceed Horizontal Limit

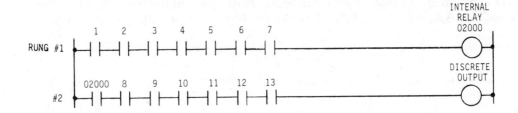

Figure 7-24b. Contacts Split into Two Rungs

NOTE: It is not necessary to split the contacts in any ratio. If the network allows 10 horizontal contacts, 10 could be placed on the top rung, and 3 could follow the N.O. contacts of the internal relay (02000) on rung 2. Needless to say, this technique would apply not only to N.O. contacts but also to N.C. or combination of N.O. and N.C. contacts as well.

Remember that the internal or "dummy" relay just used does not exist as a "real world" device that has to be hard wired but is merely a bit in the storage memory that performs the logic of a relay. In actual programming, internal control relays that do not actually exist, except in the storage memory as bits, are extensively used. The use of these internal relay equivalents is what makes the programmable controller so unique and eliminates hours and hours of hard wiring. Eliminating hard wired control relays not only eliminates installation time but also maintenance time.

PROGRAMMING RESTRICTIONS

In addition to the number of horizontal contacts on one line and the number of lines in a network or rung, the programming device does not allow for programming vertical contacts as shown in Figure 7-25.

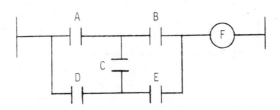

Figure 7-25. Vertical Contacts

The circuit logic says that output F can be energized by any of the following contact combinations: A,B; A,C,E; D,C,B; or D,E (Figures 7-26a, b, c, and d).

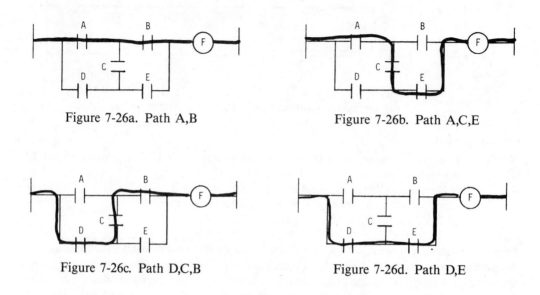

Figure 7-26a. Path A,B

Figure 7-26b. Path A,C,E

Figure 7-26c. Path D,C,B

Figure 7-26d. Path D,E

To duplicate the logic, the circuit could be programmed as shown in Figure 7-27.

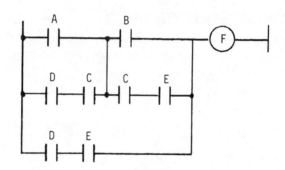

Figure 7-27. Equivalent Circuit without Vertical Contacts

This circuit maintains the circuit logic, as contact combinations: A,B; A,C,E; D,C,B; and D,E will energize output F.

Another limitation to circuit programming is the way the processor considers power flow. Flow is from left to right ONLY, and vertically UP and DOWN. The processor never allows power flow from right to left.

Normally relay logic for the circuit shown in Figure 7-28 would indicate the following possible contact combinations to energize output G: A,B,C; A,D,E; F,E; and F,D,B,C.

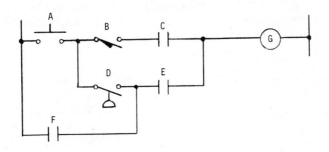

Figure 7-28. Hard Wired Circuit

If the circuit shown in Figure 7-28 was programmed into user memory as shown in Figure 7-29a, the processor would ignore contact combination F,D,B,C because it would require power flow from right to left. If combination F,D,B,C was required, the circuit would be re-programmed as shown in Figure 7-29b.

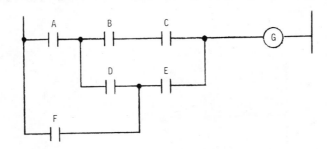

Figure 7-29a. Circuit Improperly Programmed

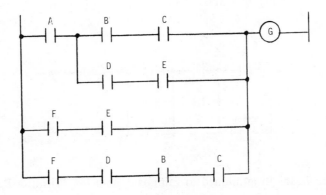

Figure 7-29b. Circuit Properly Programmed

93

The last restriction placed on programming circuits into user memory by SOME, but not all, PC's is the programming of "a branch circuit within a branch circuit" or the **nesting** of contacts. Figure 7-30a is an example of a circuit that would be referred to as having nested contacts, L and G, or "a branch within a branch." To obtain the required logic, the circuit would be programmed as shown in Figure 7-30b. The duplication of contacts J and K eliminate the nested contacts L and G.

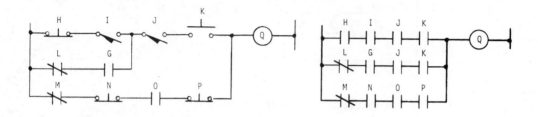

Figure 7-30a. Nested Contacts   Figure 7-30b. Programmed to Eliminate Nested Contacts

Figures 7-31a and b are another example of a circuit with a branch within a branch, in this case "branches within a branch," and how the circuit is programmed to maintain the circuit logic.

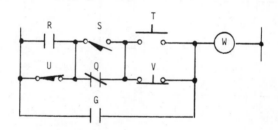

Figure 7-31a. Branches Within a Branch

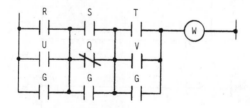

Figure 7-31b. Programmed to Eliminate Branches Within a Branch

94

Chapter Summary

The N.O. and N.C. symbols used for input devices are used differently in some applications than in others. This variation can be understood by using the relay analogy, looking at the individual bits associated with I/O devices, or using the EXAMINE ON, EXAMINE OFF true or false approach. Another approach is to accept the fact that a N.C. stop button or similar input device must be programmed using a N.O. contact symbol and have the philosophy that logic is RELATIVE TO APPLICATION and let it go at that.

Each PC has a maximum network size, or matrix, that limits the number of horizontal contacts and parallel lines for any one network or rung. The only limitation to the number of rungs (networks) is memory size.

Since the processor reads power flow horizontally from left to right ONLY and vertically either UP or DOWN, the logic of a relay circuit must be examined carefully to ensure that the logic is maintained when the circuit is programmed into the user memory.

The programming device does not allow contacts to be programmed vertically, but the logic of a ladder diagram with vertical contacts may be duplicated by adding additional contacts. Depending on the PC system, contacts may or may not be programmed as a "branch within a branch" or nested.

Review Questions

1. Define a network.
2. When a normally open (N.O.) limit switch is wired to an input module, and programmed using a N.O. contact symbol (Examine ON), the instruction will be true when: (Check all correct answers.)
   a. Power is applied and the key switch is in the run position
   b. The limit switch is closed
   c. As long as the limit switch is open
   d. Never
3. If the N.O. limit switch in Question 2 is programmed using a N.C. contact symbol (Examine OFF) the instruction will be true when: (Check all correct answers.)
   a. Power is applied and the key switch is in the run position
   b. The limit switch is closed
   c. As long as the limit switch is open
   d. Never
4. Draw and label a diagram to show how a network that requires 14 series contacts to control a discrete output can be programmed on a PC that limits series logic elements on one line to 10.

5. Indicate the logic T (True) or F (False) for the following contacts.

CONDITION OF INPUT DEVICE	PROGRAM INSTRUCTION	LOGIC TRUE - FALSE	
a.	—| |—	T	F
b.	—| |—	T	F
c.	—|/|—	T	F
d.	—|/|—	T	F

6. Draw a circuit with nested contacts.
7. Draw a circuit that retains the logic of Figure 7-A, which has a vertical contact (D), so the circuit could be programmed into a PC.

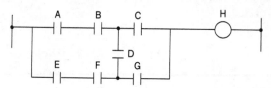

Figure 7-A

8. Power flow in a PC is considered to be: (Check all correct answers.)
   a. Up to down only
   b. Up or down only
   c. Left to right
   d. Right to left
   e. Up or down and from left to right
   f. Up, down, and from right to left
   g. Up to down only and left to right
   h. Up to down only and right to left
   i. Up or down and left to right or right to left
9. Write a program for a PC that does NOT allow nested contacts, for the hand-off auto circuit shown in Figure 7-B.

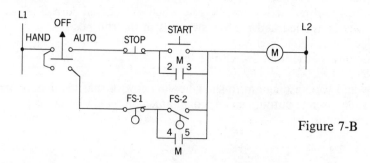

Figure 7-B

Chapter 8

Programming Devices (Programmers)

Objectives

After completing this chapter, you should be able to
- Describe the function of a programming device.
- Understand the ASCII code and parity.
- Explain the term "on line programming."
- Define baud rate.
- Understand basic programming techniques.
- Describe the FORCE ON and FORCE OFF feature and the hazards that could be associated with its use.

Programming devices, programmers, come in basically two types: desktop and hand held (Figures 8-1a and b).

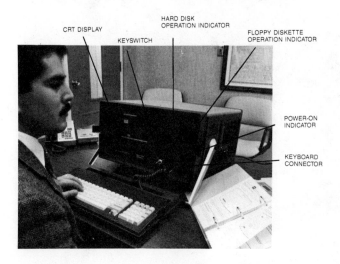

CRT DISPLAY
HARD DISK OPERATION INDICATOR
KEYSWITCH
FLOPPY DISKETTE OPERATION INDICATOR
POWER-ON INDICATOR
KEYBOARD CONNECTOR

Figure 8-1a. Desktop Programmer
(Allen-Bradley T50 Industrial Terminal)

Figure 8-1b.
Hand Held Programmer
(Courtesy of Westinghouse)

DESKTOP PROGRAMMERS

Desktop programmers like the one pictured in Figure 8-1a are designed to be portable and are built to withstand the mechanical shock associated with moving the programmer from job site to job site. Their design also allows the programmer to function in environments with electromagnetic noise, high temperatures, and humidity.

A typical programmer consists of a CRT (Cathode Ray Tube), keyboard, and the necessary electronic circuitry and operating memory for developing, modifying, and loading a program into the processor memory.

CRT

CRT's give the user a visual display of the program and range in screen or tube sizes from 5″ to 12″ measured diagonally, with the video display being black and white or green on green. CRT's are also referred to as VDT's or video display tubes. Video brightness and contrast adjustments are found either on the front or rear of the programmer. A video jack is available on the rear of most programmers for connecting an additional video monitor.

KEYBOARD

The keyboard may have raised keys, like a standard computer terminal, or have flush, sealed touchpad designed keys (Figures 8-2a and b).

Figure 8-2a. Programmer with Raised Keys
(Courtesy of Square D Company)

98

Figure 8-2b. Programmer with Flush Keys
(Courtesy of Allen-Bradley)

Keyboards will have electric symbol and special function keys, along with numeric or number keys for addressing. Many keyboards also have alphanumeric (letters and numbers) keys for report generation and other special programming functions.

The alphanumeric (letters and numbers) keys of many programming terminals generate standard ASCII characters and control codes. ASCII is an acronym for **A**merican **S**tandard **C**ode for **I**nformation **I**nterchange. The ASCII code uses different combinations of 7 bit binary (base 2) information for communication of data. The data may be communicated to a printer, tape loader, floppy, or hard disc, or be displayed on the VDT of the programmer and/or computer.

NOTE: ASCII information is also often expressed in hexadecimal (base 16). Figure 8-3 shows the 128 standard ASCII control code and character set with both the binary and hexadecimal numbering systems.

| BINARY→ | HEX→ | 000 | 001 | 010 | 011 | 100 | 101 | 110 | 111 |
		0	1	2	3	4	5	6	7
0000	0	NUL	DLE	SP	∅	@	P	\	p
0001	1	SOH	DC1	!	1	A	Q	a	q
0010	2	STX	DC2	"	2	B	R	b	r
0011	3	ETX	DC3	#	3	C	S	c	s
0100	4	EOT	DC4	$	4	D	T	d	t
0101	5	ENQ	NAK	%	5	E	U	e	u
0110	6	ACK	SYN	&	6	F	V	f	v
0111	7	BEL	ETB	'	7	G	W	g	w
1000	8	BS	CAN	(	8	H	X	h	x
1001	9	HT	EM	)	9	I	Y	i	y
1010	A	LF	SUB	*	:	J	Z	j	z
1011	B	VT	ESC	+	;	K	[	k	{
1100	C	FF	FS	,	<	L	\	l	¦
1101	D	CR	GS	-	=	M	]	m	}
1110	E	SO	RS	.	>	N	∧	n	~
1111	F	SI	US	/	?	O	—	o	DEL

(Left side vertical label: LSB — LEAST SIGNIFICANT BIT)

Figure 8-3. 128 Standard ASCII Control Code and Character Set
with Both Binary and Hexadecimal Numbering Systems

The digital or hexadecimal number is determined by first locating the vertical column where the code or character is located and then the horizontal line.

EXAMPLE: The letter A is in column 4, horizontal line 1. The binary number that will transmit the letter A is 100 0001. The hexadecimal number would be 41. The symbol # would be 010 0011 in binary and 23 in HEX.

An eighth bit is often used by programmers to provide error checking of information that is transmitted. This eighth bit is called the parity bit.

For even parity, the parity bit (eighth bit) is added to the seven bits that represent the ASCII codes and characters so that the number of 1's will always add up to an even number.

EXAMPLE: The binary number for the # symbol was 010 0011. The 1's add up to three or an odd number. By adding an eighth bit and making it a 1, the total of 1's would now be 4 or even as shown in Figure 8-4.

Figure 8-4. Parity Bit Set to 1 for Even Parity

The letter A, which was the binary number 100 0001, has two 1's or is already even. In this case the parity bit would be a 0 as shown in Figure 8-5.

Figure 8-5. Parity Bit Set to 0 for Even Parity

The ASCII control code BS (backspace) is binary number 000 1000. For even parity a 1 would be added for the parity bit as shown in Figure 8-6.

Figure 8-6. Even Parity

By checking each character or control code that is sent for an even number of 1's, transmission errors can be detected when an odd number of 1's are found.

For systems that operate on odd parity, the parity bit is used to make the total of 1's add up to an odd number.

EXAMPLE: The number 5 has binary number of 011 0101. The 1's add up to 4. The parity bit would be set to 1 making the 1's total 5 or odd. Figure 8-7 illustrates this concept.

PARITY
BIT | ASCII CODE BITS

ODD PARITY   1 |0 1 1 0 1 0 1

Figure 8-7. Parity Bit Set to 1 for Odd Parity

For systems that do not use a parity bit for error checking, the eighth bit is always a zero (0).

The electrical symbol keys that represent N.O. contacts, N.C. contacts, branch circuit start and end, coils/output, timers, counters, and so forth use what are called **OP** codes or operating codes to tell the processor what to do.

With a 16 bit word, four bits will usually be used for an OP code to tell the processor what to do, and the remaining 12 bits will be the address of where to do it. When the key for a normally open contact symbol is pushed and then followed by an address, one word of user memory is used. Four bits tell the processor to treat this as an N.O. (Examine ON) contact, and the next twelve bits will hold the address or location of the input, output or internal location that the contact is associated with.

For most PC systems, each N.O. and N.C. contact, branch start/end, and coil (output) instruction requires one word of user memory, while timers and counters will require from 2–5, depending on the PC. In reality then, when a contact is entered and addressed, one full word of user memory is used. Many programmers will display the total number of memory words used on the CRT. The total will also include any words of memory, if any, that the processor used for internal purposes.

Since the user program will require 1 full word for each contact or coil and 2 or more words for each timer/counter programmed, it is not unusual for the user memory to use more of the total memory words than the storage memory, as the storage memory only uses 1 bit of a 16 bit word to store the address or location of discrete inputs and discrete and internal outputs.

For processors with 8 bit words, contacts will normally use 2 words, while timers/counters and coils will use 4 or more words each.

Programmers that can create, modify, monitor, and load programs into user memory can also make changes to the program while the processor and driven equipment is running. This feature is often referred to as **"on line programming,"** and changing the program while the processor is running must **only** be done by persons with a complete understanding of not only the circuit operation but also the process or driven equipment as well. To prevent unauthorized "on line programming," a key switch is provided either on the programming device or on the processor. With the switch placed in the run position and the key removed, the programmer cannot be used for "on line programming." The key switch can also restrict the programmer to a monitor only mode or a program (off line) and monitor only mode. Which function(s) of the programmer that can be locked out with the key switch is (are) pretty standard but may vary from PC to PC.

PERIPHERALS
An RS-232C communication port is mounted on the back of the programmer, or on the processor, for connection to a printer for hard copy printouts of the program, storage registers, or report generation. To use a printer, the printer must be compatible with the programmer or processor.

Most newer printers can be connected for **serial** and/or **parallel** communications. In serial communications, the bits are sent one after the other or sequentially. In parallel communications, groups of bits (byte) are sent simultaneously. The rate of communication or **baud rate** between the printer and programmer will vary with the printer and/or programmer capabilities. A baud rate of 110 indicates that 10, 7 or 8 bit characters can be printed per second; 300 baud would be 30 characters per second; 1200 baud would be 120 characters per second, and so on.

Before purchasing a printer be sure and check with the local PC factory representative. There is no sense in buying a printer capable of printing 120 characters per second, 1200 baud rate, if the programmer can only communicate at 300 baud or to purchase a printer that can only print 30 characters per second, 300 baud rate, for a PC capable of 1200 baud. For printers that do not print as fast as the PC can communicate, an internal **buffer** is often used. A buffer is a temporary storage area where information can be held while the printer catches up.

To provide a dependable back-up of the program in case the memory fails or is inadvertently cleared or altered, a cassette tape recorder can be connected to the programmer, or processor, to record and/or load the user program. Data quality tapes should be used to assure accurate transfer of data.

SAMPLE PROGRAMS

Figure 8-8 shows the keyboard for the Square D Company class 8010 SY/MAX programmer. The keyboard consists of standard alphanumeric keys in typewriter format. All the keys except the CTRL and SHIFT keys generate standard **ASCII** code characters. Cursor movement control and numeric keys are located to the right of the keyboard, and a row of multiple function **soft keys** are located across the top. Soft keys are keys that perform different functions, depending on the programmer mode.

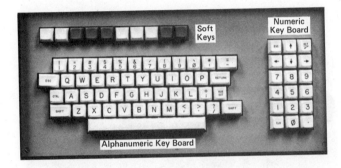

Figure 8-8. Keyboard of a Square D Company SY/MAX CRT Programmer

When the programmer is first turned on, it enters the initial mode, and the CRT display would be as indicated in Figure 8-9. The information at the bottom of the screen indicates the function of the 10 soft keys. In the initial mode, only the first 6 keys are functional. By pressing the first soft key, status, the programmer would go to the status display as indicated by Figure 8-10. From the STATUS mode the first key could be pressed again to enter the LADDER mode. From the ladder mode, either SEARCH, DELETE, or PROGRAM modes could be pressed.

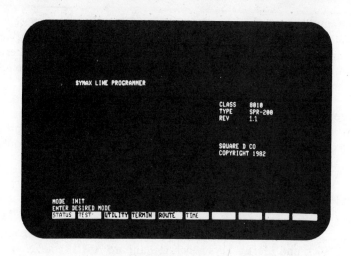

Figure 8-9. Initial Mode CRT Display
(Courtesy of Square D Company)

PRESS

The result will be:

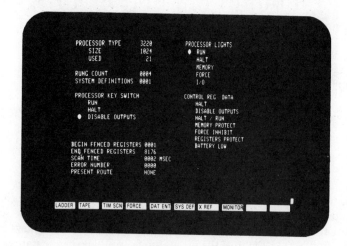

Figure 8-10. Status Mode CRT Display
(Courtesy of Square D Company)

To actually program a circuit using the programmer, the programmer is turned on. The INITIAL mode is pressed followed by the STATUS soft key, then the LADDER soft key, and then the PROGRAM soft key. The CRT would now be displaying the PROGRAM mode as shown in Figure 8-11.

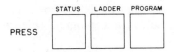

The result will be:

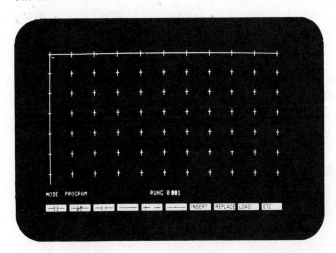

Figure 8-11. Program Mode-CRT Display
(Courtesy of Square D Company)

The soft keys now represent N.O. contacts, N.C. contacts, coil (output), space (no contact), open, and branch circuit start/close. The last 4 keys are INSERT for inserting a rung into the ladder diagram, REPLACE for replacing a rung in the ladder diagram, LOAD for loading the programmer rung(s) into memory, and ETC. When the ETC. key is pressed it changes the soft keys for programming timers/counters, and other special programming features.

By using the basic relay type instruction keys (Figure 8-12a), the circuit can be programmed as shown in Figure 8-12b.

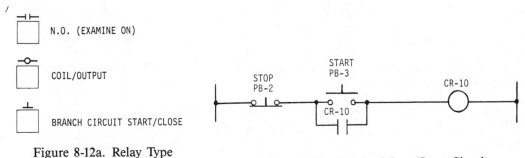

Figure 8-12a. Relay Type
Instruction Keys

Figure 8-12b. Standard Stop/Start Circuit

105

The **cursor,** which is controlled by the cursor movement keys at the top of the numeric key section of the keyboard, is used to indicate position on a rung. Initially the cursor will be at the extreme top left hand portion of the screen. This is rung 1 and the first horizontal contact position.

The first step in programming the circuit in Figure 8-12b would be to power up the programmer and press the "soft key" sequence: status, ladder, and program (Figure 8-13).

PRESS STATUS LADDER PROGRAM

Figure 8-13. Proper Keying Sequence
to Enter Program Mode

The CRT would now appear as shown in Figure 8-14. Notice the cursor is at the top left of the circuit matrix.

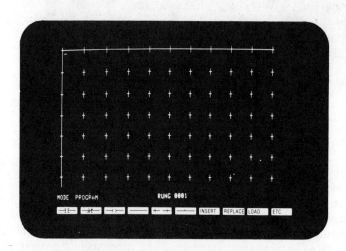

Figure 8-14. CRT Display for Program Mode
(Courtesy of Square D Company)

By pressing the key sequence shown in Figures 8-15a, b, and c, the circuit in Figure 8-12a can be programmed.

NOTE: The "soft keys" are indicated by a symbol or word **above** the key while dedicated or "hard keys" have the character **on** the key.

PRESS

The result will be:

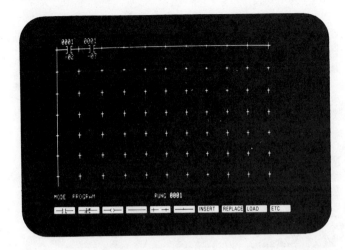

Figure 8-15a. Key Strokes for Entering and Addressing the Stop and
Start Buttons (PB-2 and 3) and the Resulting CRT Display
(Courtesy of Square D Company)

PRESS

The result will be:

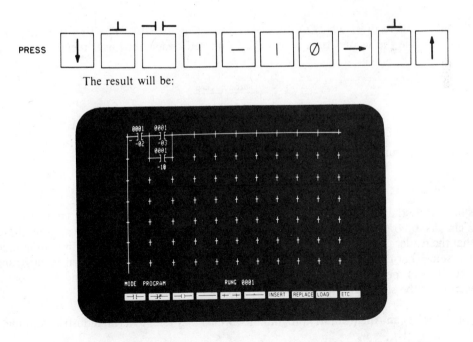

Figure 8-15b. Key Strokes for Entering and Addressing the Parallel
Holding Contacts and the Resulting CRT Display
(Courtesy of Square D Company)

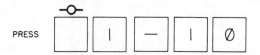

PRESS

The result will be:

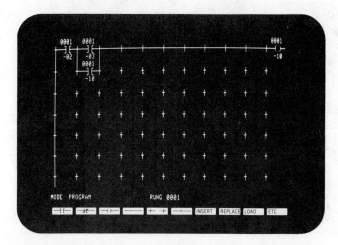

Figure 8-15c. Key Strokes for Entering and Addressing the
Output and the Resulting CRT Display
(Courtesy of Square D Company)

The final step is to press the LOAD soft key twice (Figure 8-16). The first time formats the circuit while the second time enters the formatted circuit into user memory.

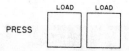

Figure 8-16. Key Sequence for Loading Rung into Memory

It is not within the scope of this text to go into great detail on programming large or complex circuits, but rather it is the purpose of this text to illustrate some simple programming so that the reader gets a **feel** for how relatively simple programming is. Each manufacturer uses a somewhat different technique, and the only way you can really learn to program a given PC is to spend time on **that** PC. While the techniques for programming differ, basic concepts are the same, and it is these **concepts** that this book covers.

Next let's look at an Allen-Bradley programmer, which they refer to as an industrial terminal.

Figure 8-17a shows the sealed touchpad keyboard used with the Allen-Bradley industrial terminal, while Figure 8-17b shows the overlays that fit over the sealed touchpads to identify the keys. Figure 8-17c illustrates how an overlay is installed on the keyboard.

108

Figure 8-17a. Sealed Touchpad Keyboard
(Courtesy of Allen-Bradley)

Figure 8-17b. Keyboard Overlays
(Courtesy of Allen-Bradley)

Figure 8-17c. Installing an Overlay
on the Keyboard
(Courtesy of Allen-Bradley)

Mounting
Notch

Mounting
Tab

Keytop Overlay

The overlay shown in Figure 8-18 is for programming and editing circuits using a T3 industrial terminal. The key groupings include numerics with force ON/OFF instructions, relay type instructions, timer/counter instructions, data manipulation, arithmetic instructions, editing instructions, control instructions, and so forth. Additional overlays are available with alphanumeric (standard typewriter format) and alphanumeric/graphic mode.

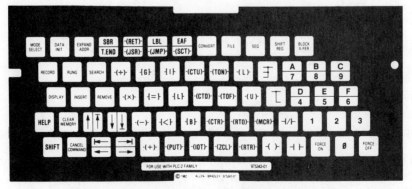

Figure 8-18. Programming and Editing Overlay
(Courtesy of Allen-Bradley)

When the programmer is connected to a Mini-PLC-2/15 and turned on, the mode select display will appear as shown in Figure 8-19. For programming a circuit, the processor mode must be selected. This is done by entering the number 11 on the keyboard. The CRT will now display an empty screen with the words START and END (Figure 8-20). The START will remain at the front of any circuit programmed, and the END, which indicates the end of the circuit, will automatically move down as contacts and/or rungs are added.

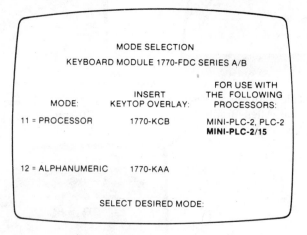

Figure 8-19. Initial CRT Display
(Courtesy of Allen-Bradley)

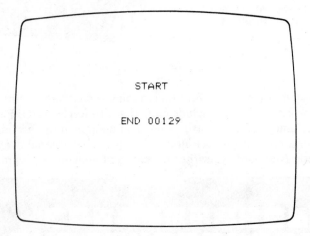

Figure 8-20. Processor Mode CRT Display

Notice that following the end statement is the number 00129. This number represents the total of memory words used. One word is for the end statement, and 128 words are for the data table (Chapter 5, Figure 5-3). If the I/O table was configured for 256 words, the number shown would be 257, 1 word for END + 256 data table words. As a program is entered, the number of additional memory words used will be added to the total (Figures 8-23a, b, c, and d).

The basic relay instructions are shown in Figure 8-21.

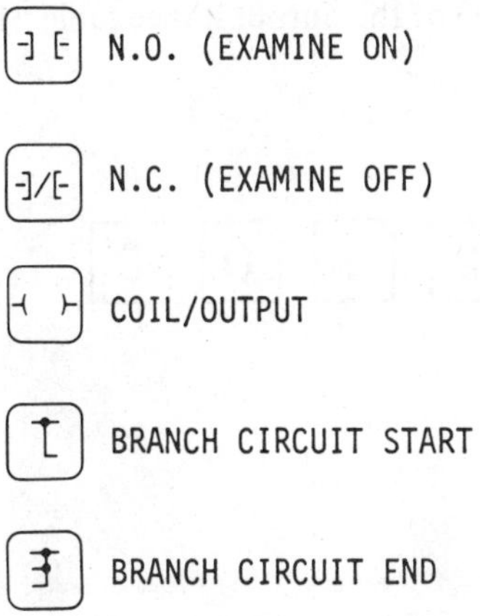

┤ ├ N.O. (EXAMINE ON)

┤/├ N.C. (EXAMINE OFF)

┥ ┝ COIL/OUTPUT

Ꮲ BRANCH CIRCUIT START

Ꮰ BRANCH CIRCUIT END

Figure 8-21. Relay Instructions

To program the circuit shown in Figure 8-22 into user memory, the key sequences shown in Figures 8-23a, b, c, and d are used.

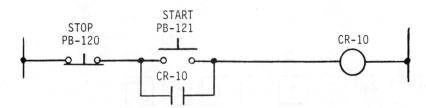

Figure 8-22. Standard Stop/Start Circuit

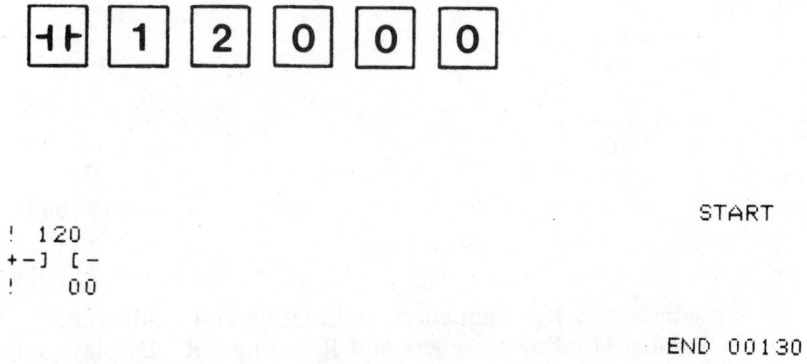

┤├ 1 2 0 0 0

```
                                                              START
!  120
+-] [-
!   00
                                                              END  00130
```

Figure 8-23a. Key Sequence for Entering and Addressing
Stop Button PB-120 and the Resulting CRT Display

111

```
                                                      START

! 120     121
+-] [-+-] [-
!  00!  00

                                           END  00132
```

Figure 8-23b. Key Sequence for Entering and Addressing
Start Button PB-121 and the Resulting CRT Display

```
LADDER DIAGRAM DUMP
                                                START
! 120     121
+-] [-+-] [-+
!  00!   00!
!     ! 010 !
!     +-] [-+
       00

                                   END  00135
```

Figure 8-23c. Key Sequence for Entering and Addressing
Parallel Holding Contacts and Resulting CRT Display

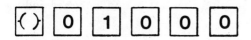

```
                                                      START
! 120     121                                                        010  !
+-] [-+-] [-+--------------------------------------------( )--+
!   00!   00!                                                        00  !
!     ! 010 !                                                            !
!     +-] [-+                                                            !
!         00                                                             !

                               END  00136
```

Figure 8-23d. Key Sequence for Entering and Addressing
the Output and Resulting CRT Display

NOTE: The number following the end statement in Figure 8-23d indicates that 136 words of the total memory have now been used. Before programming was started, the number was 129 which included 128 words for the data table plus one word for the END statement. Subtracting this total, 129, from total memory words, 136, gives us the total user memory words used for the program (7):

1 word for contact 12000
1 word for branch circuit start
1 word for contact 12100
1 word for branch circuit start
1 word for contact 01000
1 word for branch circuit end
<u>1 word for coil 01000</u>
7 Total words of user memory.

CRT DISPLAY OF CIRCUIT(S) IN USER MEMORY

Once the program has been entered into user memory, the PC is ready to control the circuit. When the processor is placed in the run mode and the circuit activated, the programmer CRT will give a visual display of the circuit condition.

Actual circuit condition is displayed in basically two ways on the CRT display. Some PC's intensify or make brighter all contacts, interconnecting lines, and coils that are passing current or have power flow, while others intensify or use reverse video only to indicate which contacts and coils have power flow. Figure 8-24a illustrates how a circuit would appear before the start button is pushed for a system that intensified contacts, interconnecting lines, and coils while Figure 8-24b shows the CRT display after the start button is pushed and the holding contacts close.

Figure 8-24a. CRT Display Prior to Start Button Being Depressed

Figure 8-24b. CRT Display After Start Button
is Depressed and Holding Contacts Close

Figure 8-25a illustrates how a display using reverse video would look before the start button is pushed while Figure 8-25b shows the display after the start button is depressed and the holding contacts close.

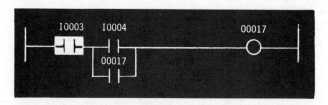

Figure 8-25a. Reverse Video Display Prior to Start Button Being Depressed

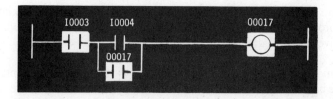

Figure 8-25b. Reverse Video Display After Start Button
is Depressed and Holding Contacts Close

No matter which method is used, this feature of desktop programmers is a powerful troubleshooting aid. By viewing the display on the CRT, the maintenance electrician can determine which contacts are closed and which outputs are turned on. A word of caution though, an output coil that is intensified only indicates that the output module circuit is on. It does not guarantee that the actual discrete output device is ON. However, if the discrete output device is all right, it will be ON any time the output module circuit is ON.

As a further troubleshooting aid, most PC's are designed so any input or output contact (N.O. or N.C.) or output coil can be **FORCED ON** or **OFF**. This feature allows the operator to FORCE or make a contact or coil go ON regardless of actual input device status or circuit logic. Likewise, contacts can be FORCED or turned OFF, again regardless of actual input device status or circuit logic.

CAUTION: The FORCE ON-FORCE OFF feature should never be used except by personnel who completely understand the circuit **AND** the process machinery or driven equipment. An understanding of the potential effect that forcing a given contact or coil will have on machine operation is essential if hazardous and/or destructive operation is to be avoided.

HAND HELD PROGRAMMERS

Unlike the desktop programmer that can display a complete circuit network on the CRT, most hand held programmers have limited display capabilities. Some hand held programmers can display a rung with up to four horizontal lines, while others can only display one line or one element at a time. The display will either be LED's or liquid crystals. Figure 8-26 shows a Modicon Micro 84 programmer with liquid crystal display.

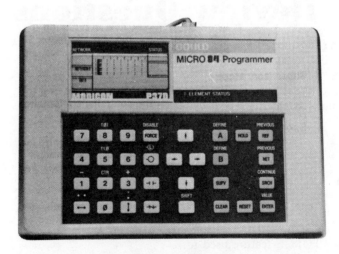

Figure 8-26. Modicon Micro 84 Programmer with Liquid Crystal Display
(Courtesy of Gould, Inc.)

Most of the hand held programmers can perform many of the programming functions of the desktop programmer, and some have cassette tape loading capabilities.

Hand held programmers are well suited for installations that require constant changes in circuit requirements since they are light-weight (approx. 2 pounds), portable, and ruggedly constructed.

Hand helds are, of course, cheaper than desktop programmers which makes them an affordable troubleshooting tool. While it takes more time to go through the program one contact or rung at a time, the difference in cost may make the extra time acceptable.

Chapter Summary

The programming device or programmer is used to enter, modify, and monitor the user program. The program (ladder diagram) is entered by pushing keys on the keyboard in a prescribed sequence with the results being displayed on either the CRT of a desktop programmer or with an LED or liquid crystal display for a hand held programmer. The visual display can also be used as a troubleshooting aid to test the circuit prior to entry into user memory or after the circuit is entered into memory and is operational. Contacts and coils are either intensified or displayed in reverse video to indicate power flow. From the programming device, contacts and coils can be forced on or off while the circuit is operational. The FORCE ON, FORCE OFF capability should be **restricted** to personnel who have a **complete** understanding of the circuit and the driven equipment. Programming any PC is not difficult, but time must be spent to become familiar with the PC and its programming techniques.

Review Questions

1. Explain the function of a programming device.
2. What is a CRT, and what is its function?
3. What does "on line programming" mean?
4. Define baud rate.
   Add the parity bit to the seven ASCII code bits below to make even parity.
5. Parity Bit   ASCII Code Bits
      ____   010 0011
6. Parity Bit   ASCII Code Bits
      ____   110 1001
7. Parity Bit   ASCII Code Bits
      ____   000 1001
8. What is a buffer?
9. What is the function of the cursor?
10. What is the FORCE feature used for?
11. T   F   Timers and counters use words of memory, but contacts, coils, and branch start instructions do not.

Chapter 9

Small Programmable Controllers

Objectives

After completing this chapter, you should have the knowledge to
- Categorize programmable controllers by I/O size.
- Define basic Boolean terms: AND, OR, and NOT.
- Interpret truth tables.
- Use Boolean functions to write simple programs.

SMALL PCs

What's small? "So small it fits in a shoe box" that's how one manufacturer describes their small PC. How small are they? Figure 9-1 shows a Gould PC0085. Notice the Number 2 pencil used to show relative size. Small in this case is 5-¾" wide X 6" high and 4" deep.

Figure 9-1. Gould PC0085 Small Programmable Controller
(Courtesy of Gould Inc.)

Depending on whose definition you use, PC's can be grouped by I/O size.

A PC users magazine used the following I/O sizes to categorize PC's.

> Micro—Up to 64 I/O
> Small—65 to 128 I/O
> Medium—129 to 892 I/O
> Large—More than 892 I/O

The terms micro, small, mini, and compact are used interchangeably by the PC manufacturers. While there isn't an agreement on what to call this new line of miniature PC's, they all have several things in common; they are small, flexible, modular, and inexpensive.

While they are great for replacing existing relays and relay control circuits, like their big brothers, these PC's also offer timer/counter functions, math functions, data compare, data transfer, shift registers, sequencers, etc.

While not every small PC offers all these functions, you can no doubt find one that will meet your specific needs.

A typical small PC would have 800–1000 words of memory. Memory type would be CMOS RAM with a lithium battery for back-up. Optional storage or back-up could be UVPROM, EPROM, or EEPROM. Communication to a cassette recorder or IBM compatible computer would enable the user to store and/or to develop a user program.

While some of the small PC's are programmed using relay ladder logic, the majority seem to be programmed using Boolean Algebra.

BOOLEAN ALGEBRA

Boolean algebra? Don't let the terms Boolean or algebra scare you off at this point. Once you understand some Boolean concepts you will find that programming in Boolean is not only fun, but is fast, too.

Boolean algebra dates back to 1854 when mathematician George Boole developed his mathematical system of logic. In a logic problem there are only two states: true or false.

Boolean algebra is a unique system that differs from regular high school and/or college algebra. In the Boolean system there are only two digits: 0 and 1. Every number must consist of only ones (1) and zeroes (0), and there are no fractional numbers.

Relay ladder diagrams can be thought of as logic problems because relay contacts can only have two states: closed (true) 1 or open (false) 0.

To better understand the concept, consider the circuit in Figure 9-2.

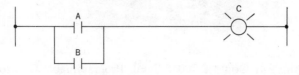

Figure 9-2. Basic OR Relationship

118

With contacts A and B wired in parallel, closing either contact A **OR** B will turn on lamp C. This type of circuit arrangement, where either A OR B can turn on C, is referred to as **OR** logic.

The Boolean equation would be A + B = C. The plus sign (+) sign indicates OR logic.

The formula says: If A **OR** B is true, then C is true. A truth table for the equation will further illustrate and clarify the Boolean concept (Figure 9-3).

NOTE: A 0 represents an open or OFF condition, while a 1 represents a closed or ON condition.

Figure 9-3. OR Truth Table

CONTACT	CONTACT	LAMP	
A	B	C	A + B = C
0	0	0	0 + 0 = 0
0	1	1	0 + 1 = 1
1	0	1	1 + 0 = 1
1	1	1	1 + 1 = 1

For series contacts as shown in Figure 9-4, a logic statement would say: If contacts A **AND** B are closed, C will light.

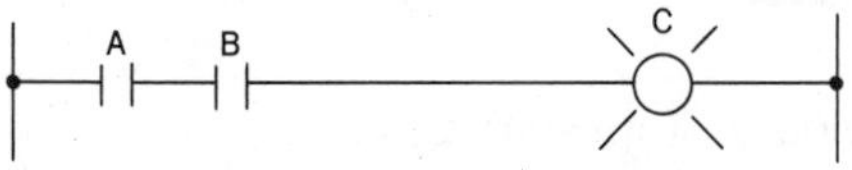

Figure 9-4. Basic AND Relationship

This circuit configuration, where both A AND B must be closed to turn on C, is referred to as **AND** logic. The Boolean formula would be written A X B = C. The multiplication sign (X) indicates AND logic.

NOTE: The AND function can also be expressed A · B = C or AB = C.

The formula says: If A **AND** B are true, then C is true. A truth table for the equation is illustrated in Figure 9-5.

Figure 9-5. AND Truth Table

CONTACT	CONTACT	LAMP	
A	B	C	A X B = C
0	0	0	0 X 0 = 0
0	1	0	0 X 1 = 0
1	0	0	1 X 0 = 0
1	1	1	1 X 1 = 1

Next, look at the simple circuit programmed earlier in Chapter 7 (Figure 9-6).

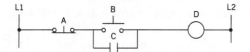

Figure 9-6. Standard Stop/Start Circuit

This circuit contains both AND logic and OR logic. The logic says: If A is closed **AND** either B **OR** C are closed, D will light. The Boolean formula would be A X (B + C) = D. Figure 9-7 shows the truth table and equations.

CONTACT	CONTACT	CONTACT	LOAD	
A	B	C	D	A X (B + C) = D
0	0	0	0	0 X (0 + 0) = 0
0	0	1	0	0 X (0 + 1) = 0
0	1	0	0	0 X (1 + 0) = 0
0	1	1	0	0 X (1 + 1) = 0
1	0	0	0	1 X (0 + 0) = 0
1	0	1	1	1 X (0 + 1) = 1
1	1	0	1	1 X (1 + 0) = 1
1	1	1	1	1 X (1 + 1) = 1

Figure 9-7. Truth Table for Standard Stop/Start Circuit

The next function that you will need to understand is the **NOT** function. The NOT function is like the Examine OFF instruction discussed in Chapter 7.

The NOT function acts like a set of N.C. contacts. Consider the circuit in Figure 9-8.

Figure 9-8. Basic NOT Relationship

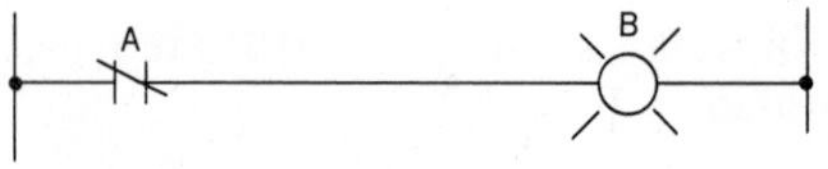

As long as A remains closed (NOT open) then lamp B will be ON.

Another approach would be to think of the **NOT** function as being True when the device or address that it represents is NOT ON or set to 0.

120

A truth table for a NOT function is shown in Figure 9-9.

CONTACT	LAMP
A	B
0	1
1	0

Figure 9-9. NOT Truth Table

Combining the NOT function with the AND function is illustrated in the circuit and truth table shown in Figure 9-10.

CONTACT	CONTACT	LAMP
A	B	C
0	0	0
0	1	0
1	0	1
1	1	0

Figure 9-10. Circuit and Truth Table for AND and NOT Functions

Combining the NOT function with the OR function is illustrated in the circuit and truth table shown in Figure 9-11.

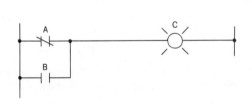

CONTACT	CONTACT	LAMP
A	B	C
0	0	1
0	1	1
1	0	0
1	1	1

Figure 9-11. Circuit and Truth Table for NOT and OR Functions

Combining the AND, OR, and NOT functions are illustrated in the circuit and truth table shown in Figure 9-12.

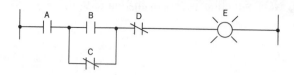

CONTACT	CONTACT	CONTACT	CONTACT	LAMP
A	B	C	D	E
0	0	0	0	0
0	0	0	1	0
0	0	1	0	0
0	0	1	1	0
0	1	0	0	0
0	1	0	1	0
0	1	1	0	0
0	1	1	1	0
1	0	0	0	1
1	0	0	1	0
1	0	1	0	0
1	0	1	1	0
1	1	0	0	1
1	1	0	1	0
1	1	1	0	1
1	1	1	1	0

Figure 9-12. Circuit and Truth Table Combining AND, OR and NOT Functions

PROGRAMMING IN BOOLEAN

Since the Gould PC0085 method of programming seems to be somewhat typical of small PC's that are programmed in Boolean, the following discussion will be based on the Gould Model PC0085.

Excluding timers, counters, and special function keys, programming ladder logic requires only the following 7 keys:

1.	STR	(Store)
2.	AND	(AND function)
3.	OR	(OR function)
4.	NOT	(NOT function)
5.	OUT	(Output)
6.	ENTER	(Enter instruction)
7.	NEXT ADRS	(Next address)

Figure 9-13 shows a simple line of ladder logic and the keystrokes required for programming.

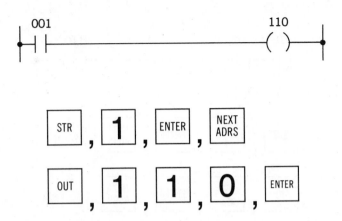

Figure 9-13. Programming Simple Line of Ladder Logic

Programming is started by pushing the STR (Store) key followed by up to a 3 digit address.

NOTE: It is not necessary to enter zeros that precede numbers in addresses. The address 001 would require that only a 1 be pushed. The address 050 would require that only a 5 and a 0 be pushed.

The STR key followed by the number 1 would enter a N.O. contact addressed 001.

Pushing the ENTER and NEXT ADRS (next address) keys load the N.O. contact 001 and readies the processor for the next instruction.

Pressing the OUT (output) key and then the three digit number 110 will give us an output with address 110. Pressing the enter key will load the rung of logic.

NOTE: Because the OUTPUT was the last element in the program, only the ENTER key was needed to load the information. If an additional rung or rungs were to be programmed, then both the ENTER and NEXT ADRS keys would need to be used.

Once the programming has been completed, the processor could be placed in the monitor mode (run mode) to verify the program.

Figure 9-14 shows two contacts in series, or an AND function, and the keystrokes required for programming.

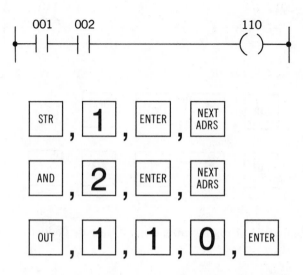

Figure 9-14. Programming AND Function

The keystrokes STR, 1, again gave us a N.O. contact 001. The keystrokes AND, 2, put an additional N.O. contact 002 in series with N.O. contact 001. The keystrokes OUT, 1, 1, 0, provided OUTPUT 110 to complete the line of logic.

To program N.C. contacts or a NOT function, the STR key is followed by the NOT key as shown in Figure 9-15.

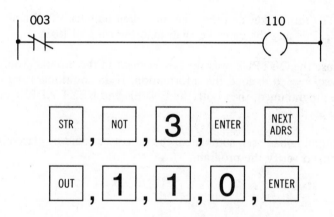

Figure 9-15. Programming NOT Function

Figure 9-16 shows a circuit with a N.O. contact 001 in parallel or ORed with a N.C. or NOT contact 003 controlling OUTPUT 110. The keystrokes to program the circuit are also shown.

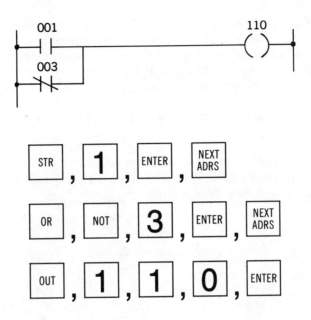

Figure 9-16. Programming OR Function

For more complex circuits where AND and OR functions are combined, the processor uses a temporary file or stack system to store information as the circuit is being developed.

Each time the STR key is pushed the information that follows is put into the temporary file or stack. Information enters the stack from the bottom and pushes previous information up. Figure 9-17 illustrates this concept.

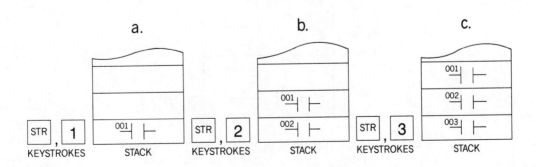

Figure 9-17. Entering Information Into Temporary File (Stack)

NOTE: Because it is understood that the ENTER key and the NEXT ADDRESS key must be used during each step of program development, those keystrokes will no longer be shown.

When the first STR (STR, 1) was entered at (a), N.O. contacts 001 were placed on the bottom of the stack. When the second STR (STR, 2) was entered at (b), N.O. contacts 002 are inserted into the bottom of the stack and contacts 001 move up one place. At position (c), the third STR command was entered and N.O. contacts 003 enter the bottom of the stack and the previous contacts moved up.

At any point in the program the contacts can be retrieved in inverse order, or last-in-first-out (LIFO).

The keystrokes AND, STR would take the last contact entered (003) and AND it with the next contact up in the stack (002). Figure 9-18 shows what the stack now looks like after the AND, STR commands have been entered.

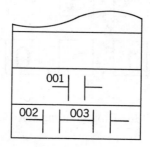

Figure 9-18. LIFO Stack After AND, STR Command

Had we entered the keystrokes OR, STR the contacts 003 and 002 would have been ORed as shown in Figure 9-19.

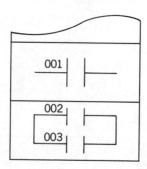

Figure 9-19. LIFO Stack After OR, STR Command

126

With the contacts as shown in Figure 9-19, an AND, STR, series of keystrokes would AND the ORed contacts 002 and 003 with contacts 001 as shown in Figure 9-20.

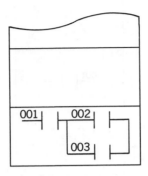

Figure 9-20. LIFO Stack After AND, STR Command

Different STR commands would have yielded different results on the original stack in Figure 9-17. Figures 9-21a, b, and c show some alternative strokes and what the results would be.

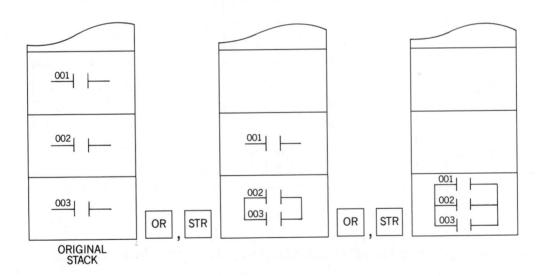

Figure 9-21a. Stack After Two Consecutive OR, STR Commands

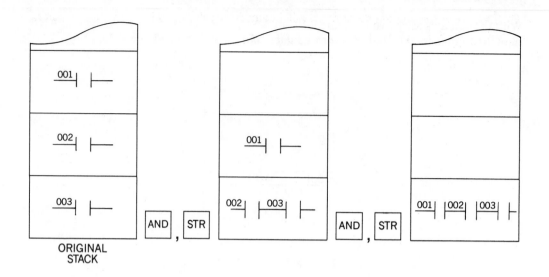

Figure 9-21b. Stack After Two Consecutive AND, STR Commands

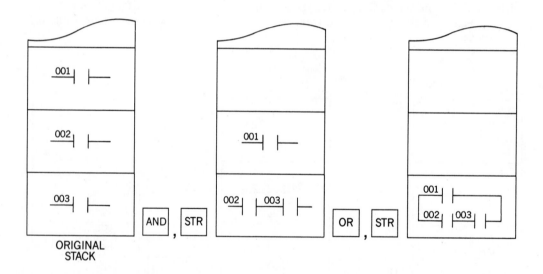

Figure 9-21c. Original LIFO Stack After AND, STR and OR, STR Commands

The circuits shown in Figures 9-21a, b, and c could have been programmed without first developing a stack of three contacts.

This method was used to demonstrate a concept, and was used for illustration only.

The keystrokes to program the three parallel contacts, as shown in Figure 9-21a, would be:

> STR 1
> OR 2
> OR 3

To program the contacts in series, as shown in Figure 9-21b, the following keystrokes would be used.

> STR 1
> AND 2
> AND 3

The following keystrokes could be used to program the parallel series combination circuit shown in Figure 9-21c.

> STR 2
> AND 3
> OR 1

To complete the circuits shown in Figures 9-21a, b, and c the OUTPUT key would be pressed and a 3 digit address entered. While not all small PC's allow for parallel outputs, Gould's PC0085 does. Figure 9-22 is an example of parallel outputs and the keystrokes required for programming.

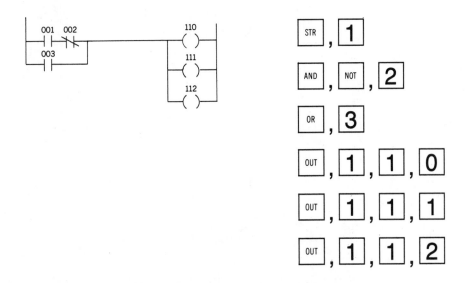

Figure 9-22. Programming Parallel Outputs

NOTE: Remember that ENTER and NEXT ADDRESS would follow each step during programming and are not being shown only for simplicity.

As stated earlier in this chapter, programming in Boolean is fun and fast. Notice that no branch START or branch END commands are needed which really speeds programming time.

To further illustrate the programming techniques and the LIFO (last in, first out) stack concept, Figure 9-23 shows a series parallel combination circuit with the keystrokes and the temporary stack results.

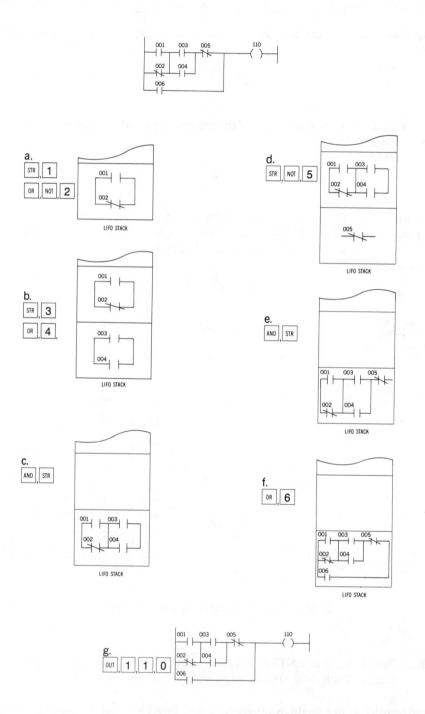

Figure 9-23. Programming a Combination Circuit and the Resulting LIFO Stack

130

The keystrokes in (a) connect contacts 001 and 002 in parallel and places them in the LIFO stack.

The keystrokes in (b) connect contacts 003 and 004 in parallel and enters this data into the stack. This moves contacts 001 and 002 up to make room for contacts 003 and 004 on the bottom.

Step (c) brings contacts 001 and 002 back down to the bottom of the stack and ANDs them with contacts 003 and 004.

The STR, NOT, 5, keystrokes in (d) moves the ANDed contacts up in the stack and a N.C. NOT contact 005 moves into the bottom of the stack.

The AND, STR command at step (e) recalls the contacts to the bottom of the stack and ANDs them with the NOT contact 005.

Step (f) recalls the contacts from the LIFO stack and ORs them with contact 006.

Finally, in step (g) the output is added with keystrokes: OUT, 110.

TIMERS

Programming timers with the Gould PC0085 is quick and quite simple.

Figure 9-24 shows a simple ON delay timer circuit and the necessary keystrokes.

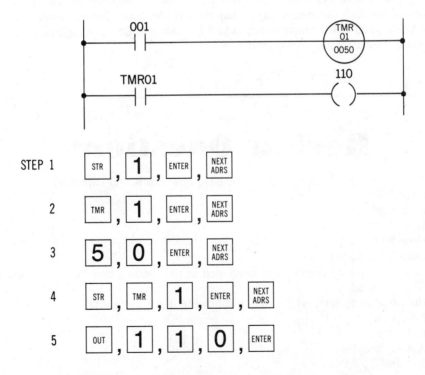

Figure 9-24. Programming Simple On-Delay Timer

131

Step 1 enters the contact to control the timer (N.O. 001).

Step 2 enters a timer and identifies it as timer number 1.

Step 3 enters the preset for timer 1. The time base for PC0085 timers is .1 seconds. Timer 1 in this illustration would be 5 seconds (50 X .1 sec = 5 seconds).

Step 4 enters a set of timer contacts that will close 5 seconds after timer 1 (TMR 1) is energized.

Step 5 completes the circuit with output 110. Output 110 is controlled by timer contacts and will be delayed by 5 seconds after input device 001 is closed.

The Gould PC0085 has many other features and program capabilities, as do the other small PC's on the market today. The examples and discussion in this chapter provide only an introduction to programming in Boolean, but should give you a basic understanding to enable you to read the user manuals for any small PC and get started programming.

Chapter Summary

Programmable controllers can be grouped by I/O size. Generally, any PC with up to 128 I/O is categorized as a small PC. The small PC's are in fact small physically, but offer many of the features of larger, more expensive PC's. Many of the small PC's are programmed using limited Boolean statements rather than relay ladder logic. Typical Boolean functions are AND, OR, and NOT. Programming a small PC that uses Boolean logic is fast and quite simple once the basics are understood.

Review Questions

1. To be categorized as large, a PC must have how much I/O capacity?
   a. Over 128
   b. Over 528
   c. Over 712
   d. Over 892
   e. None of the above
2. T   F   Boolean algebra was developed in the 1960's when PC's were first being used.
3. In the Boolean system, which of the following digits are used?
   a. 1, 2, 3 and 4
   b. 1, 3, and 5
   c. 0, 1, 2, and 4
   d. 0, 1 and 2
   e. None of the above

4. What Boolean relationship do contacts connected in parallel have?
   a. AND
   b. OR
   c. NOT
   d. IF
5. In a Boolean equation a plus sign indicates which type of logic?
   a. AND
   b. OR
   c. NOT
   d. IF
6. A multiply sign (X) indicates what type of Boolean logic?
   a. AND
   b. OR
   c. NOT
   d. IF
7. The NOT function acts in a circuit like a set of:
   a. N.O. contacts
   b. N.C. contacts
   c. Timed contacts
   d. Intermittent contacts
8. Using the Gould PC0085 programming format, list the keystrokes to program the circuit shown below.

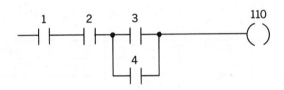

9. List the keystrokes necessary to program the following circuit using a Gould PC0085.

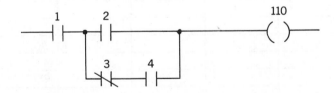

10. What are the keystrokes required to enter the following circuit into Gould PC0085 memory?

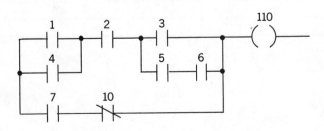

133

11. Complete the truth table for the circuit shown below.

CONTACT 1	CONTACT 2	CONTACT 3	OUTPUT 110

12. Complete a truth table for the circuit shown in Question 9.

CONTACT 1	CONTACT 2	CONTACT 3	CONTACT 4	OUTPUT 110

Chapter 10

Latching and Master Control Relays

Objectives

After completing this chapter, you should be able to
- Write a program using a latching relay.
- Understand the term *retentive*.
- Write a program using a master control relay.
- Understand the importance of a safety circuit.

LATCHING RELAYS

Before discussing how latching relays are programmed with a PC, a review of the traditional "hard-wired" latching relays may be helpful.

Latching relays are used where it is necessary for contacts to stay open and/or closed even though the coil is only energized for a short time (40 milliseconds). Latching relays are also used for master stop-start applications where the coil of the latching relay is momentarily energized to close contacts, and the contacts remain closed even though the relay coil is no longer energized.

Latching relays normally use two coils for latch and unlatch with mechanical linkage that holds the relay in the latched or closed position. When the unlatch coil is energized, the coil action disengages the mechanical latch and allows the relay to open.

Figure 10-1 is the wiring diagram for a mechanical latching relay.

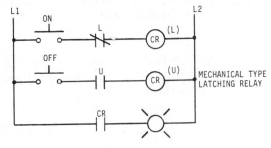

Figure 10-1. Mechanical Latching Relay

When the ON button is pushed, the latch coil energizes and opens the N.C. L contacts and closes the U contacts. Opening the N.C. L contacts de-energizes the latch (L) coil. The length of time the ON button was pushed, energizing the latch coil, which opened the N.C. L contacts and de-energizes the latch coil, required only a fraction of a second. During the short time the latch coil energized, it closed the N.O. CR contacts completing the circuit to the lamp. CR contacts remain closed even though the latch coil de-energizes because of the mechanical latch mechanism. To open the mechanically latched contacts to turn OFF the light requires that the OFF pushbutton be pushed. The U contacts in the unlatch coil circuit are now closed, and pushing the OFF button energizes the unlatch coil, which in turn closes the N.C. L contacts and opens the U contacts, which de-energizes the unlatch coil. For the brief instant that the unlatch coil was energized, it released the mechanically latched CR contacts so they could open and turn OFF the light.

Mechanical latching relays can be replaced by programming bits in the storage memory of the PC as internal latching relays. Like the dummy relays discussed earlier, the programmed internal latching relays do not exist as "real world" devices but can perform all the logic of an actual latching relay.

Programmed latch/unlatch relays, like their physical "real world" counterparts, are **"retentive"** on power failure. That is to say, if the relay is latched, it will remain latched if power is lost and then restored.

Figure 10-2 shows latch and unlatch rungs as they would be programmed on an Allen-Bradley PLC-2/30.

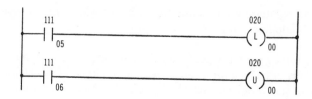

Figure 10-2. Programmed Latch and Unlatch Rungs

Both the latch (L) and unlatch (U) coils have the same address (02000). The address, which is bit 00 of word 020, will be set to 1 or ON when the latch rung is true, input 11105 closed, and will be cleared to 0 or OFF when the unlatch rung is true, input 11106 closed. Like normal latching relays, only a momentary closure of input device 11105 will latch the output coil (L) 02000, and the output will remain latched or ON until the unlatch coil (U) rung is true by closing input 11106.

N.O. and N.C. contacts programmed after the latch/unlatch rungs and given the same address as the latch/unlatch coils will perform the same functions as N.O. and N.C. contacts of a standard "real world" latching relay (Figure 10-3).

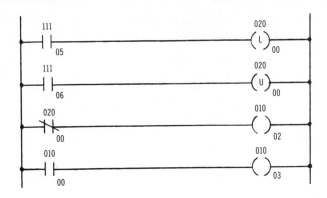

Figure 10-3. Programmed Latching Circuit

Output 01002 is ON and output 01003 is OFF when 02000 is not latched.

Output 01002 will go OFF and 01003 will come ON when 02000 is latched.

Normally, an internal storage bit (dummy relay) will be used for the latch and unlatch address, rather than an actual discrete output address. If a discrete output address is used, the output, once latched, will remain ON even if programmed after an open master control relay (MCR) rung. When an internal storage bit is used for the latch and unlatch address, the bit will still be retentive, but will turn off if programmed after a MCR rung that is open. Programmed MCR's are covered next in this chapter.

Figure 10-4 shows how a latch circuit would appear on the CRT of a General Electric Series Six programmer.

Figure 10-4. Latching Circuit as Displayed on a
General Electric Series Six Programmer CRT

Output 00001-L (latch) comes ON when input I0001 goes true or ON. Any N.O. contact in the circuit addressed 00001 would close and remain closed or latched until input I0002 (reset latch) goes true or turns ON. When I2 goes on, output 00001-U (unlatch) is cleared to 0, and any N.O. contacts addressed 00001 would open or go false. Likewise, any N.C. contacts addressed 00001 would go closed or become true.

MASTER CONTROL RELAYS

In standard relay control systems, a master control is often used to control power to the entire circuit or to just selected rungs. This allows selected rungs or the whole circuit to be de-energized by turning off the MASTER CONTROL RELAY. Figure 10-5 shows a typical "hard wired" master control relay that controls power for the whole circuit.

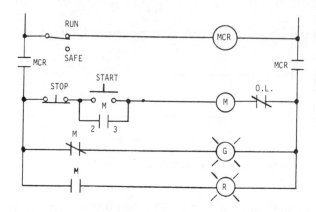

Figure 10-5. Hard Wired Master Control Relay

Master control relays are often used with circuits that have off delay timers so the circuit can be shut down completely **without** waiting for the timers to time out.

With PC's, a master control relay (MCR) can be programmed to control an entire circuit or to just control selected rungs of a circuit. When the MCR is programmed as shown in Figure 10-6, any rungs that follow can only become energized if MCR is energized or true. When MCR is true, the outputs in the rungs that follow are controlled by the logic programmed for each rung. If MCR is de-energized, the rungs below the MCR **cannot** energize even if the programmed logic for each rung is TRUE.

WARNING: Once latched (ON), discrete devices addressed as latched relays will remain ON even when the MCR is false (de-energized). To avoid this problem, use an internal storage bit for the latch/unlatch address and Examine On and Off instructions with the same address to control discrete devices.

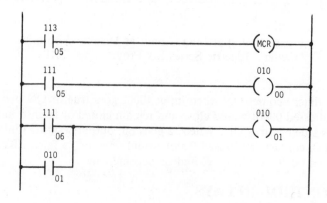

Figure 10-6. Programmed MCR

138

Figure 10-7 shows a programmed circuit with two MCR's. By adding the second MCR, the rungs between the two MCR's, rungs 3 and 4, are controlled by the first MCR and contact 03-01. The second MCR, rung 5, is independent of the first and controls rung 6 below it.

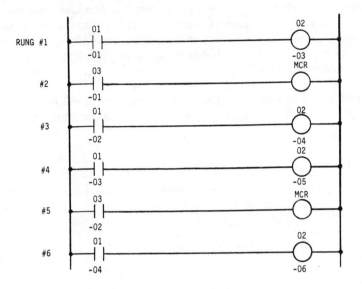

Figure 10-7. Using Multiple MCR's

To end the control of one MCR, a second MCR can be programmed **unconditionally**, with no contacts or logic preceding the MCR, as shown in Figure 10-8. Any additional rungs below the second MCR will function normally with the output being energized when the programmed logic is true, regardless of whether the first MCR is energized or not.

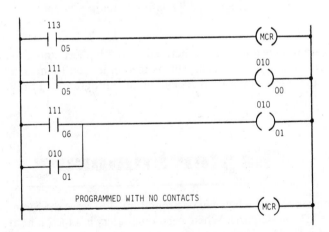

Figure 10-8. Unconditionally Programmed MCR

NOTE: Latching relays will still be retentive even when programmed in a rung following a MCR.

SAFETY CIRCUIT

The National Electrical Manufacturing Association (NEMA) standards for programmable controllers recommends that consideration be given to the use of emergency stop functions which are independent of the programmable controller. The standard reads in part: "When the operator is exposed to the machinery, such as loading or unloading a machine tool, or where the machine cycles automatically, consideration should be given to the use of an electro-mechanical override or other redundant means, independent of the controller, for starting or interrupting the cycle."

Figure 10-9 shows how a control relay (CR) and safe-run switch could be added to interrupt lines 1 and 2 to the discrete output devices of an automatic machine or process.

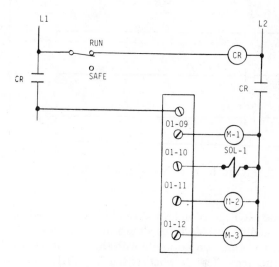

Figure 10-9. Safety Circuit

While all PC's are designed and manufactured to the highest standards and quality, the safety circuit should always be added. This is recommended by most PC manufacturers to ensure maximum safety rather than depending on a programmed MCR or latching relay alone.

Chapter Summary

Latching and master control relays can be programmed to serve the same control functions as their "real world" counterparts. Where personnel safety is a factor, a safety circuit should be added rather than depending on a latching or master control relay alone.

Review Questions

1. Will both programmed and "real world" latching relays, if latched, remain latched if power is removed (lost) and then restored?
2. Latching relays are normally used when it is necessary for
   a. Contacts to open and/or close only while the coil is energized.
   b. Contacts to open and/or close every 30 seconds.
   c. Contacts to stay open and/or closed even though the coil is only energized a short time.
   d. None of the above.
3. Explain why a master control relay (MCR) is often used with OFF DELAY timers.
4. Master control relays can be used to control
   a. Selected circuit rungs (networks)
   b. Entire circuits
   c. Individual contacts within a rung (network)
   d. All of the above
5. Define the term "unconditionally."
6. When using PC's, the National Electrical Manufacturing Association (NEMA) recommends that consideration be given to stop functions independent of the PC. Explain briefly why this recommendation was made.

Chapter 11

Programming Timers

Objectives

After completing this chapter, you should have the knowledge to
- Describe how pneumatic time delay relays work.
- Write a program using ON delay and OFF delay timers.
- Describe the difference between an ON delay timer and a retentive timer.
- Explain how to extend the time range of timers by *cascading* timers.

PNEUMATIC TIMERS (GENERAL)

To fully understand how a PC can be programmed to replace pneumatic time delay relays, both the basic pneumatic time delay relay and the standard symbols used need to be understood.

Figure 11-1 shows a complete Allen-Bradley pneumatic timing relay and Figure 11-2 shows a cutaway view of the contact and timing mechanism.

Figure 11-1. Pneumatic Timing Relay
(Courtesy of Allen-Bradley)

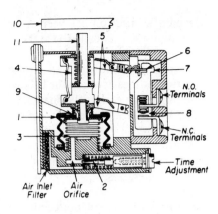

Figure 11-2. Cutaway View of Contact Unit and Timing Mechanism
(Courtesy of Allen-Bradley)

For the timer to time when power is applied (coil energized), the solenoid unit (coil, core piece, and armature) is mounted so the natural weight of the armature (10) pushes down on the operating plunger (11) which causes the bellows (1) and bellows spring (3) to collapse the bellows and dispel the air in the bellows out through the release valve (9). When the coil is energized, the armature is attracted magnetically to the pole pieces and lifts up and off the bellows assembly. Air now comes in through the air inlet filter, past the needle valve (2), filling the bellows with air. The incoming air expands the bellows upward, pushing on the timing mechanism plunger (4). As the plunger rises, it causes the over-center toggle mechanism (5) to move the snap action toggle blade (6) upward which picks up the push plate (7) that carries the movable contacts (8) to open the N.C. contact and close the N.O. contact. The time it takes for the bellows to fill with air and activate the contact mechanism is controlled by adjusting the needle valve in the air orifice. The valve is adjusted with a screwdriver as shown in Figure 11-3, and counterclockwise rotation will move the needle valve further into the air orifice and restrict air flow into the bellows, slowing the air flow and increasing the time it takes for the bellows to expand and operate the contact mechanism. Clockwise adjustment of the needle valve will decrease the time it takes the bellows to fill with air and activate the contacts after the armature has been lifted off the bellows mechanism.

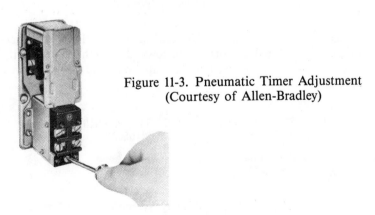

Figure 11-3. Pneumatic Timer Adjustment
(Courtesy of Allen-Bradley)

When the contact action is delayed after the coil has been energized and the armature is lifted up and off the bellows mechanism, it is called **ON DELAY**.

When the coil of an ON DELAY timer is de-energized, the armature drops down, pushing on the operating plunger, which in turn pushes down on the bellows, expelling the air through the release valve. The bellows moving down causes the snap action toggle blade to instantaneously snap the N.C. contact CLOSED and the N.O. contact OPEN.

To summarize the ON DELAY relay, it is clear that the delay in contacts opening and closing occurs when the coil is energized or ON, and the contacts go back to their NORMAL condition instantly when the coil de-energizes.

Figure 11-4 shows a pneumatic timer with the solenoid unit mounted for ON DELAY.

Figure 11-4. ON Delay Timer
(Courtesy of Allen-Bradley)

Figure 11-5 illustrates the electrical symbols used to indicate ON DELAY contacts.

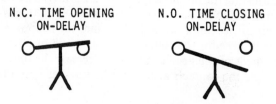

Figure 11-5. ON Delay Symbols

The arrowhead indicates that movement is UP, and since ON DELAY contacts can only time after the armature has lifted UP off the bellows, this method of identifying timed contacts is easy to remember. The other common method of identifying timed contacts is shown in Figure 11-6.

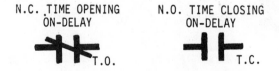

Figure 11-6. ON Delay Symbols

Remember that NORMAL for contacts is how they are, open or closed, with the coil of the relay de-energized.

For a pneumatic timer to time when the power is removed from the relay coil or **OFF DELAY**, the solenoid unit is mounted as shown in Figure 11-7. With a spring holding the armature up, no weight is applied to the bellows assembly, so the bellows are filled with air and in a fully extended position.

Figure 11-7. OFF Delay Timer
(Courtesy of Allen-Bradley)

NOTE: Compare Figure 11-4, the ON DELAY, with Figure 11-7 to clearly see the different mounting of the solenoid assemblies.

When the coil is energized and the armature moves down, the armature pushes on the operating plunger which in turn pushes on the bellows assembly, and all air is immediately forced out of the bellows through the release valve. This causes the snap-action contact assembly to instantly OPEN the N.C. contact and CLOSE the N.O. contact. The contacts will stay in this configuration as long as the coil is energized and the armature is holding the bellows mechanism down or compressed.

When the relay coil is de-energized or turned OFF, the spring on the armature lifts it up and off the operating plunger which allows the bellows to start to fill with air. The N.C. contact will remain OPEN, and the N.O. contact will remain CLOSED until the bellows has filled with enough air to activate the snap-action contact mechanism. When the contact mechanism has been activated, the N.C. contacts go CLOSED and the N.O. contacts go OPEN.

Figure 11-8 shows the electrical symbols for OFF DELAY contacts.

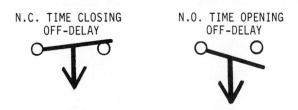

N.C. TIME CLOSING
OFF-DELAY

N.O. TIME OPENING
OFF-DELAY

Figure 11-8. OFF Delay Symbols

To avoid confusion when reading electrical drawings with OFF DELAY contacts, it must be remembered that NORMAL means after the coil has been de-energized, turned OFF, and the time set for the timer has ELAPSED. The other symbols used for OFF DELAY contacts are shown in Figure 11-9.

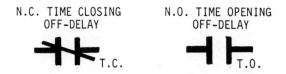

Figure 11-9. OFF Delay Symbols

Figure 11-10 compares both types of symbols used for ON DELAY and OFF DELAY timing relays.

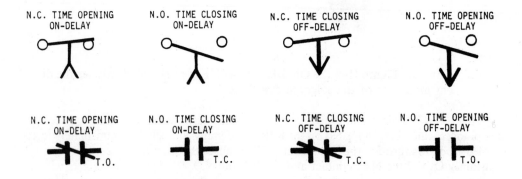

Figure 11-10. ON and OFF Delay Symbols

By reviewing the two types of symbols commonly used in motor control diagrams, a person should have no trouble determining the type of timing relay (ON DELAY or OFF DELAY) used and what is normal (open or closed) for the timed contacts.

The basic pneumatic timing relay is designed so that additional instantaneous contacts may be added as shown in Figure 11-11. The instantaneous contacts operate when the coil is energized or de-energized independent of the timing mechanism. Figure 11-12 shows the electrical symbol for contacts with an asterisk (*) which is sometimes used to indicate instantaneous contacts of a timing relay.

Figure 11-11. Adding Instantaneous Contacts
(Courtesy of Allen-Bradley)

N.O. INSTANTANEOUS CONTACTS   N.C. INSTANTANEOUS CONTACTS
TIME DELAY RELAY   TIME DELAY RELAY

Figure 11-12. Instantaneous Contact Symbols

Figure 11-13a shows a simple light circuit controlled by an ON DELAY timer set for 5 seconds. The amount of delay is written by the timer coil on the diagram for understanding and for troubleshooting. When S^1 is closed (Figure 11-13b), the coil of the pneumatic timer energizes, lifts the armature up and off the bellows, and the timing starts. Figure 11-13c shows the circuit after 3 seconds have elapsed, not enough time for the timer to time out, with the lamp circuit still open. After 5 seconds have elapsed (Figure 11-13d), the N.O. time closing contacts CLOSE, and the lamp lights. As long as S^1 remains closed, the timer coil will be energized, and the timed contacts will stay closed. When S^1 is opened (Figure 11-13e), the coil circuit is broken, and the coil de-energizes, which in turn opens the timed contacts turning off the lamp. The timed contacts will open the instant the coil de-energizes since they are timed only when power is applied to the coil.

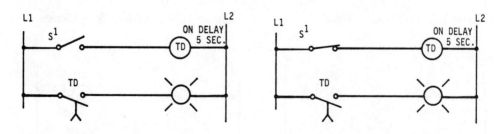

Figure 11-13a. ON Delay Timer Circuit   Figure 11-13b. Instant S^1 is Closed

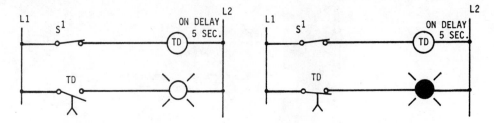

Figure 11-13c. 3 Seconds After S¹ is Closed   Figure 11-13d. 5 Seconds After S¹ is Closed

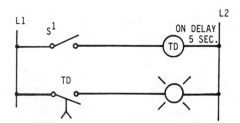

Figure 11-13e. Instant S¹ is Opened

Figure 11-14a shows the same circuit with an OFF DELAY timer. When S¹ is closed (Figure 11-14b), the TD coil energizes drawing the armature down and compressing the bellows which causes the N.O. OFF DELAY contacts to go CLOSED instantly, and the lamp will light. When S¹ is opened (Figure 11-14c), the TD coil is de-energized, the spring loaded armature is lifted up and off the bellows, and the 5 second timing begins. Figure 11-14d shows the circuit after 3 seconds have elapsed. The lamp will remain energized until the full 5 seconds have elapsed and the N.O. contacts time out and open (Figure 11-14e).

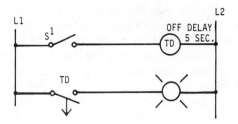

Figure 11-14a. OFF Delay Timer Circuit

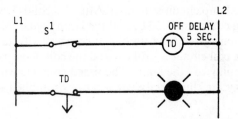

Figure 11-14b. Instant S¹ is Closed

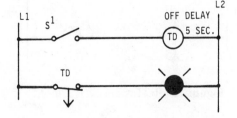

Figure 11-14c. Instant S¹ is Opened

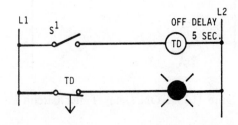

Figure 11-14d. 3 Seconds After S¹ is Opened

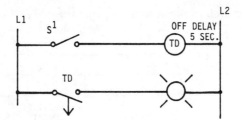

Figure 11-14e. 5 Seconds After S¹ is Opened

Instead of the bellows assembly, like the pneumatic time delay relay, PC timers use internal solid-state circuitry (clock) for timing intervals or timing base.

The various PC manufacturers use varying approaches for the actual programming of timers. Four methods which are typical for most PC's will be discussed.

Because it is an easy transition from pneumatic timer concepts to programming timers concepts, the Allen-Bradley approach to programming timers will be discussed first.

ALLEN-BRADLEY TIMERS

The amount of time for which a timer is set is called the **preset** time. As the timer is activated and starts timing, time is **accumulated** until the preset time is reached. When the accumulated time equals the preset time, the contacts are activated.

Allen-Bradley timers use two words of memory. One word stores the preset time, and another word stores the accumulated time. When a timer is programmed, the operator must enter an address, a time base, and a preset time (PR) as indicated in Figure 11-15.

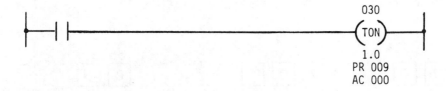

Figure 11-15. Programming an Allen-Bradley ON DELAY (TON) Timer

The accumulated time (AC) will display the actual accumulating time as the timer is timing. The timer address is used to identify the first word that will be used for the timer from the data table. On a standard 128 word data table, the first word set aside for timers and counters is word 030. If word 030 is used for the timer address, the first 12 bits, bits 00-07 and 10-13, of word 030 will store the accumulated time of timer 030 in Binary Coded Decimal (BCD) format (Figure 11-16).

NOTE: Allen-Bradley word and bit addresses use the octal numbering system.

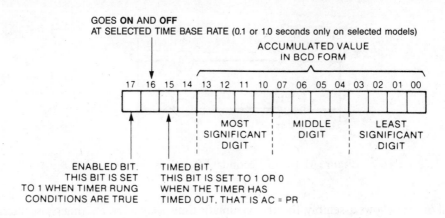

GOES **ON** AND **OFF**
AT SELECTED TIME BASE RATE (0.1 or 1.0 seconds only on selected models)

Figure 11-16. Time Counted Using BCD (12 Bits)

Three of the last four bits are used as status bits. Bit 15 is set to 1 or 0 depending on whether the timer is ON or OFF DELAY (TON or TOF). Bit 16 on some systems goes ON and OFF (1 and 0) at the rate of the time base. Bit 17 is set to 1 or turned on whenever the timer is energized. Bit 15 acts like timed contacts, while bit 17 acts as instantaneous contacts.

As discussed in Chapter 5, Allen-Bradley automatically sets aside the word numbered 100 higher than the timer for storing the preset value. If word 030 is addressed as a timer, word 130 is automatically used to store the preset value. Addressing word 031 as a timer would cause the preset value to be stored in word 131.

The time base can be seconds, tenths of a second, or hundredths of a second. Time bases are selected by entering numerical information into the processor with the programmer (industrial terminal). For a time base of seconds, 10 is entered, and 1.0 will be displayed on the CRT. To set the time base for tenths of a second, 01 is entered, and the CRT displays 0.1. Entering 00 gives hundredths of a second or .01 (Figure 11-17).

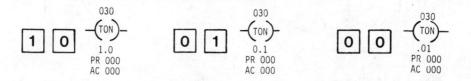

Figure 11-17. Time Bases

Next, the preset value is entered. The processor converts the decimal number(s) entered into BCD format and stores the information in the word that is 100 higher than the timer address.

Figure 11-18a shows the CRT display for an ON DELAY timer (TON) addressed 030, and Figure 11-18b shows word 130 and the preset value of timer 030 stored in BCD.

Figure 11-18a. Programmed ON Delay Timer (TON)

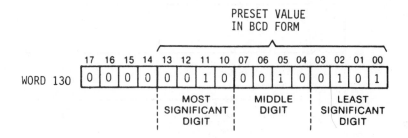

Figure 11-18b. Preset Value Stored in Word 130

By using BCD format and 12 bits, the largest number or preset time that can be stored is 999. The other 4 unused bits of word 130 could be used as internal storage.

As indicated earlier, when the timer is energized, the accumulated time (value) will be stored in the first 12 bits of word 030 in BCD format. The CRT, however, will display the accumulated time in decimal numbers.

To better understand how the timer works, let's program the circuit shown in Figure 11-19a using Allen-Bradley format (Figure 11-19b).

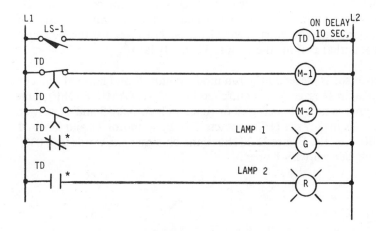

Figure 11-19a. ON Delay Timing Circuit

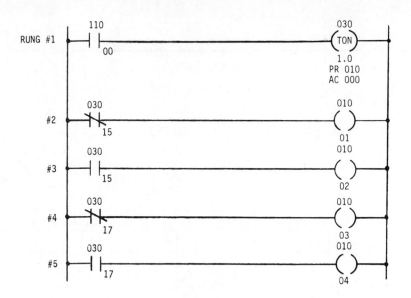

Figure 11-19b. Timing Circuit Programmed Using Allen-Bradley Format

First the limit switch 1 (LS-1) is programmed as a N.O. contact (EXAMINE ON) and given an input address of 11000, word 110-bit 00. Next an ON DELAY timer (TON) is programmed and addressed 030. The time base is set for seconds, 1.0, and the preset value of 10 seconds is entered, 010.

Figure 11-19a, motor #1 is controlled by a N.C. time opening contact. To program the equivalent, a N.C. (EXAMINE OFF) contact is used and addressed 03015. Remember that bit 15 of word 030 is a timed bit and will be set to 1 or turn ON when the timer has timed out. For motor #2, which is controlled by a N.O. time closing contact, a N.O. (EXAMINE ON) contact is programmed and also addressed 03015. Remembering that bit 17 is turned ON or set to 1 any time the timer is energized, bit 17 is used for the N.O. and N.C. instantaneous contacts that control the lamps (Figure 11-19b).

When the processor is placed in the run mode, motor #1 and lamp #1 will come ON. Both motor #1 and lamp #1 (green) are controlled by N.C. (EXAMINE OFF) contacts. As LS-1 is open, the timer is not yet energized or timed out, so bits 15 and 17 of word 030 are set to 0 or OFF. This makes the N.C. contacts TRUE, so motor #1 and lamp #1, bits 01 and 03 of word 010, are ON. Figure 11-20 shows the bit status for words 010, 030, 110, and 130 with power on, but with LS-1 open.

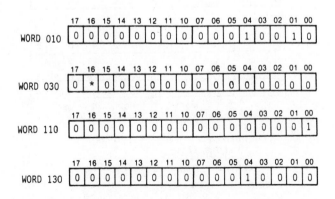

	17	16	15	14	13	12	11	10	07	06	05	04	03	02	01	00
WORD 010	0	0	0	0	0	0	0	0	0	0	0	0	1	0	1	0

	17	16	15	14	13	12	11	10	07	06	05	04	03	02	01	00
WORD 030	0	*	0	0	0	0	0	0	0	0	0	0	0	0	0	0

* On some Allen-Bradley systems bit 16 will go ON and OFF at the set time base. For a time base of 1 second, bit 16 would be ON (1) for 1 second then OFF (0) for 1 second, ON (1) for 1 second, etc. This occurs any time power is applied to the system regardless of whether or not the timer is energized.

	17	16	15	14	13	12	11	10	07	06	05	04	03	02	01	00
WORD 110	0	0	0	0	0	0	0	0	0	0	0	0	0	0	0	0

	17	16	15	14	13	12	11	10	07	06	05	04	03	02	01	00
WORD 130	0	0	0	0	0	0	0	0	0	0	0	1	0	0	0	0

Figure 11-20. Bit Status of Words 010, 030, 110, and 130

When LS-1 is closed, bit 00 of word 110 is set to 1, the timer rung is true, and the timer energizes and starts to time. With the timer energized, bit 17 of word 030 is set to 1 which makes the EXAMINE OFF (N.C.) contact in rung 4 go false and the EXAMINE ON (N.O.) contact in rung 5 go true. This turns OFF lamp #1 and turns ON lamp #2.

Figure 11-21 shows the bit status for words 010, 030, 110, and 130 at the instant the timer is energized.

Figure 11-21. Bit Status of Words 010, 030, 110, and 130
the Instant the Timer Rung is TRUE

As the timer counts in 1 second intervals, the accumulated time is stored in the first 12 bits of word 030. Figure 11-22 shows the bit status for word 030 as it counts from 000, energized, to 010, timed out.

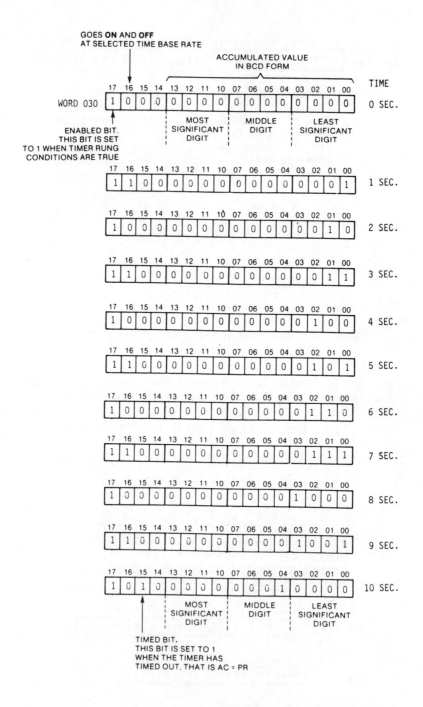

Figure 11-22. Bit Status as Accumulated Time Goes from 000 to 010

When the timer times out (accumulated time equals the preset time) timed bit 15 of word 030 is set to 1 or ON. Bit 15 set to 1 makes rung 2 (Figure 11-19b) EXAMINE OFF (N.C.) contact go false, turning OFF motor #1. Rung 3 EXAMINE ON (N.O.) contacts are now true, and motor #2, bit 02-word 010, is turned ON.

Initially when the preset time of 010 was stored in word 130, the processor on each scan would compare the accumulated time in word 030 with the preset value in 130. On the scan that the accumulated value (time) equaled the preset value (time), the processor updated bit 15 of word 030 and turned it ON.

Like its physical counterpart, any time power is removed from the timer, the timer will reset to 000.

Due to scan times, most PC's have a timer accuracy of + or − 10m seconds. The timer accuracy decreases as scan time increases. If the accuracy is critical, beyond + or − 10m seconds, most PC's have special programming techniques to compensate for scan times.

Figure 11-23a shows a circuit with a pneumatic OFF DELAY timer, and Figure 11-23b shows how the same circuit would be programmed and addressed.

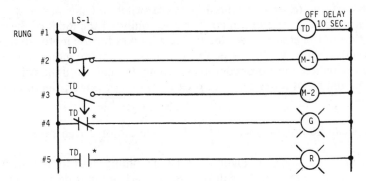

Figure 11-23a. OFF Delay Timing Circuit

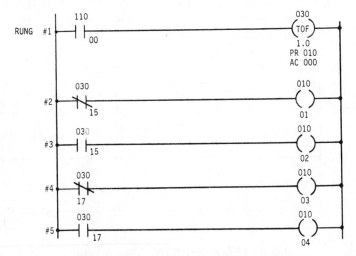

Figure 11-23b. Timing Circuit Programmed Using Allen-Bradley Format

155

When the processor is turned ON only rungs 2 and 4 are true, so only motor #1 and lamp #1 are ON.

When LS-1 (11000) closes, rung 1 goes true, the OFF DELAY timer TOF-030 is energized, and bits 15 and 17 are set to 1. This makes rungs 2 and 4 go false and rungs 3 and 5 go true, so motor #1 and lamp #1 go OFF, and motor #2 and lamp #2 go ON. Since this is an OFF DELAY timer, the circuit will stay this way as long as input device 11000 remains closed.

When input 11000 opens, OFF DELAY timer (TOF) 030 will de-energize and start to time. At the instant the timer was de-energized, bit 17 was set to 0, so rung 4 went true and rung 5 went false. Bit 15 will continue to be set to 1 or ON until the accumulated value is equal to the preset value. When the values are equal, bit 15 is set to 0 and rung 3 goes false while rung 2 goes true.

Again like its pneumatic relay counterpart, the OFF DELAY timer will time out when power is removed and reset to 000 whenever power is applied.

PC's also offer a timer that will replace the standard motor driven timer. A typical motor driven timer consists of shaft mounted cam(s) that are driven by a synchronous motor. Rotating cam(s) activate (open-close) limit or micro switches. Once power is applied, the motor starts turning the shaft and cam(s). The positioning of the lobes of the cam(s) and the gear reduction of the motor determine the time it takes for the motor to turn the cam far enough to activate the switches. If power is removed from the motor, the shaft stops. When power is re-applied, the motor continues turning the shaft until the switches are activated. When the timing of a device is not reset on a loss of power, the timing is said to be **retentive.**

Retentive timers can be programmed that will replace motor driven timers.

A retentive timer (RTO) is shown in Figure 11-24a and a timing chart for the circuit is shown in Figure 11-24b. The retentive timer will start to time when input device 11306 is closed. If the input device is opened after 3 seconds, the timer accumulated value stays at 003. When input 11306 is closed again, the timer picks up the time at 3 seconds and continues timing. When the accumulated value equals the preset value 009, bit 15 of word 052 is set to 1, and output 01004 is turned ON.

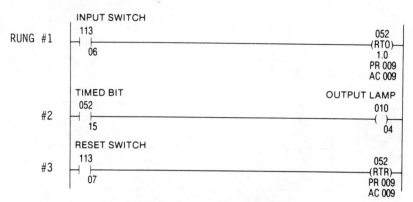

Figure 11-24a. Retentive Timer Circuit
(Courtesy of Allen-Bradley)

156

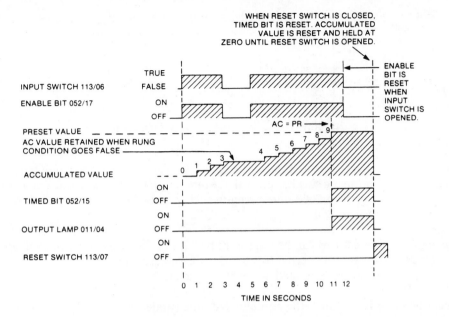

Figure 11-24b. Timing Chart
(Courtesy of Allen-Bradley)

Since the retentive timer does not reset to 000 when the timer is de-energized, a reset rung must be added. Rung 3 of Figure 11-24a illustrates how this is done. The rung consists of a N.O. input device such as a limit switch or pushbutton addressed 11307 and a retentive timer reset (RTR). The retentive timer reset is given the same addresss 052 as the retentive timer (RTO). When input device 11307 closes, RTR resets the accumulated value in word 052 and the timed bit 15 to 0. After resetting RTO, input device 11307 is opened again. If input 11306 is still closed, the retentive timer will start timing again, but if input 11306 is open, RTO will remain reset to 000 until input 11306 closes. When input 11306 closes, RTO will start to time.

When programming timers, whether they are ON DELAY (TON), OFF DELAY (TOF), or RETENTIVE (RTO), there is no limit, except for memory size, to the number of N.O. and N.C. timed and instantaneous contacts that can be programmed for any one timer.

WESTINGHOUSE TIMERS

Look now at how Westinghouse timers are programmed.

The timer for a Westinghouse PC-700 or 900 system consists of two input circuits, a timing block, and an output coil (Figure 11-25).

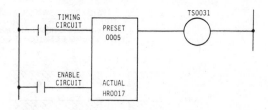

Figure 11-25. Westinghouse Timer Format

The timing and enable circuits control the timer. Both circuits must be conducting for the timer to time. If the timing circuit is opened, the timer stops timing, but the actual or accumulated time is not reset to 0000 but is retentive. If the enable circuit is opened, the timer automatically resets to 0000.

The amount of time for which the timer is to be set, or its preset time, is entered, and then a holding register (HR) is selected to store the accumulated time as the timer is counting. The preset time can be from 0001 to 9999, four digit BCD.

When the timer is activated, the accumulated time (value) in the holding register is compared to the preset time. When the accumulated value (time) equals the preset value (time), the output coil is energized. Any time that the enable circuit is opened, the value in the holding register is reset to 0000, and the output is de-energized.

The time base, either seconds or tenths of a second, is determined when the timer output coil is addressed. If the address is preceded by a TS, the time base will be seconds. If TT is entered prior to the address, the time base will be tenths of a second.

Figure 11-26 shows an ON DELAY timer, number TS0014, with the timing and enable circuits both controlled by input device IN0001. The preset time is 0005 or 5 seconds. The actual or accumulated value will be stored in holding register 1 (HR0001). Below the timer circuit is a timing chart that indicates the status of the output (TS0014) for different times in relationship to the input device (IN0001). The holding register (HR0001) shows the accumulated time when the timer is timing, or, if the enable circuit is opened, it shows that the accumulated time was reset and held at 000.

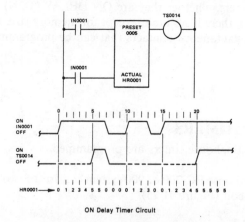

ON Delay Timer Circuit

Figure 11-26. Timing Circuit and Timing Chart
(Courtesy of Westinghouse)

158

At time 0, IN0001 is closed, turning ON both the timing and enable circuits, and the timer starts to time. If IN0001 is left closed, after 5 seconds the accumulated value in HR0001 will equal the preset value, and output TS0014 will energize as indicated at time 5. Notice also that the accumulated value in holding register HR0001 counted to five and then stopped. The accumulated value can never exceed the preset value.

Like a pneumatic ON DELAY timer, if power is removed, the timer resets to 0, and output TS0014 is de-energized. If the input device IN0001 is again closed at time 10, HR0001 starts to accumulate time. If the input device is opened at time 13, HR0001 is reset to 0000, and the output cannot energize because the accumulated value in HR0001 never equaled the preset value. At time 15 the input device is closed and left closed. When the value in HR0001 equals the preset value (time 20), output TS-0014 is again energized and would remain energized until IN0001 is again opened.

Normally open and normally closed contacts addressed TS0014 could be programmed to control discrete or internal outputs.

Figure 11-27a shows a circuit for controlling two lights, and Figure 11-27b shows how the circuit would be programmed.

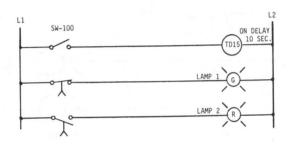

Figure 11-27a. ON Delay Timing Circuit

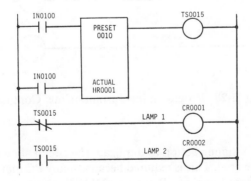

Figure 11-27b. Timing Circuit Programmed Using Westinghouse Format

NOTE: CR is the **mnemonic code** or designation Westinghouse uses for discrete inputs. IN is the mnemonic designation for discrete inputs. TT and TS are also mnemonic designations or codes.

When the PC was turned ON, CR0001 (lamp #1) would light through the N.C. time opening contacts of timer TS0015. When input device IN0100 is closed, the timer starts to time.
When the accumulated value in HR0001 equals the preset value, output TS0015 energizes,
the N.C. time opening contacts open turning OFF CR0001 (lamp #1), the N.O. time closing
contacts close, and CR0002 (lamp #2) turns ON. The lamps would remain in this condition
until input device IN0100 is again opened and the accumulated value in HR0001 was reset
to 0000 and TS0015 is de-energized.

For instantaneous contacts, a logic coil (dummy relay CR0257) controlled by a set of IN0100
contacts can be used as shown in Figure 11-28. N.O. and N.C. contacts of logic coil CR0257
are used to control discrete outputs CR0003 and CR0004.

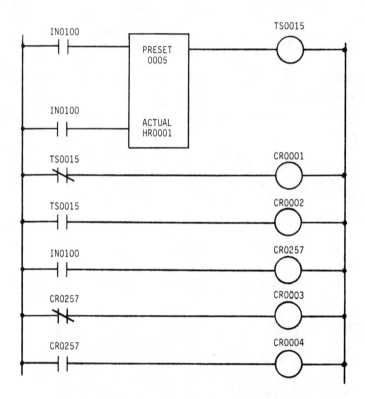

Figure 11-28. Programming Instantaneous Contacts

To program OFF DELAY timing action to duplicate the circuit shown in Figure 11-29, an
ON DELAY timer will be used. This is required because many PC manufacturers only have
dedicated ON DELAY timer functions. Before an ON DELAY timer is programmed to perform an OFF delay function, consider the operation of Figure 11-29 and the corresponding
timing chart.

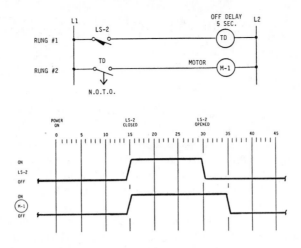

Figure 11-29. OFF Delay Timer

With LS-2 OPEN, the OFF delay timer is not energized, even with power applied to the circuit, and M-1 is not running. When LS-2 is CLOSED (time 15), the OFF delay timer is energized and the N.O.T.O. contacts in rung 2 close and M-1 is energized or ON as shown in the timing chart (Figure 11-29). When LS-2 is OPENED (time 30), the closed N.O.T.O. contacts in rung 2 remain CLOSED for 5 seconds then go OPEN (time 35) and turn OFF M-1.

To try and duplicate this circuit operation, an ON DELAY timer can be programmed as shown in Figure 11-30.

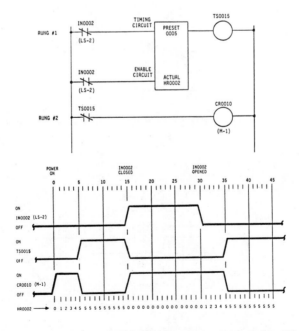

Figure 11-30. ON Delay Timer Programmed for OFF Delay Timing Action

161

The normally OPEN input contacts IN0002 (LS-2) and the normally OPEN time opening TD contacts are programmed N.C. (EXAMINE OFF) as shown in Figure 11-30. By using IN0002 N.C. (EXAMINE OFF) contacts in both the timing and enable circuits, the timer will energize and time out when power is first applied to the circuit and open the N.C. TS0015 contacts that control CR0010.

When the input device LS-2 (IN0002) is closed, the EXAMINE OFF contacts go FALSE (OPEN) and the timer is reset to 0000, the N.C. timer contacts TS0015 close, and CR0010 is energized. When the input device opens again, the N.C. (EXAMINE OFF) contacts to the timer are true, and the timer starts to time. When the accumulated value in holding register 0002 (HR0002) equals the preset value, 5 seconds, coil TS0015 energizes. When coil TS0015 energizes, the N.C. contacts to CR0010 open, and CR0010 is de-energized.

The difference with the programmed OFF delay circuit is evidenced by the timing chart (Figure 11-30). Notice that when power is applied, CR0010 (M-1) in rung 2 turns ON, even though IN002 (LS-2) is open. With IN002 programmed EXAMINE OFF, the timer will time when power is applied and the TS0015 EXAMINE OFF contacts in rung 2 will be TRUE or ON. The timer will time until the accumulated value equals the preset (5 seconds). Now that the accumulated value equals the preset, coil TS0015 is energized and the EXAMINE OFF contacts in rung 2 go false and CR0010 turns off. From this point on (time 5) this circuit will duplicate the pneumatic timing circuit in Figure 11-29 as can be seen by comparing the timing charts of each circuit as LS-2 is closed (time 15) and then opened (time 30).

Obviously, the problem with this programmed circuit is the fact that CR0010 (M-1) is energized for 5 seconds when power is first applied to the circuit. Not only could this cause sequence problems, but also could represent a serious safety hazard. By reprogramming the circuit as shown in Figure 11-31, we can accurately duplicate the timing action of the circuit in Figure 11-29.

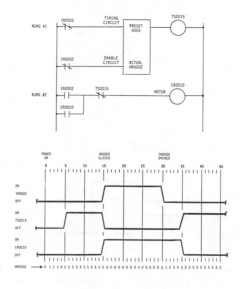

Figure 11-31. OFF Delay Timer Circuit With Timing Chart

162

N.C. (EXAMINE OFF) contacts IN0002 are again programmed for the N.O. input device LS-2 in the timing and enable circuits of the timer (rung 1) and an additional N.O. IN0002 contact is programmed in the rung 2. This N.O. IN0002 contact in rung 2 prevents CR0010 from energizing when power is applied. With the input device open, both the timing and enable circuits of timer TS0015 are true. The timer is activated and times until the accumulated value (HR0002) equals the preset value. CR0010 is further prevented from energizing by the now open TS0015 contacts. The timer will stay activated and timed out (AC = PR) until the input device is closed.

At time 15, IN0002 (LS-2) is closed which resets the timer. With TS0015 being de-energized, CR0010 can now energize through the now closed IN0002 contacts and the N.C. TS0015 contacts. CR0010 N.O. holding contacts also close. At time 30, IN0002 (LS-2) opens, causing the timer to activate. CR0010 remains energized by its holding contacts and the N.C. TS0015 contacts. When the timer times out at time 35, coil TS0015 energizes, which opens the N.C. TS0015 contacts in the CR0010 circuit, and CR0010 drops out. By not initially energizing CR0010 when power was applied, but only energizing it when IN0002 (LS-2) closed and keeping CR0010 energized 5 seconds after IN0002 opened, exactly duplicates the OFF DELAY timer circuit in Figure 11-29.

To further illustrate how to program OFF delay circuits by using ON delay timers, compare the circuits and timing charts in Figures 11-32 and 11-33 using both N.O.T.O. and N.C.T.C. contacts.

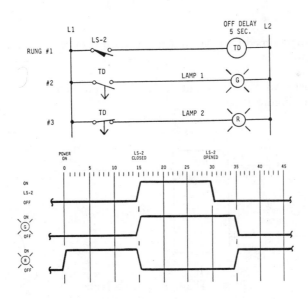

Figure 11-32. OFF Delay Timing Circuit With N.O.T.O. and N.C.T.C. Contacts

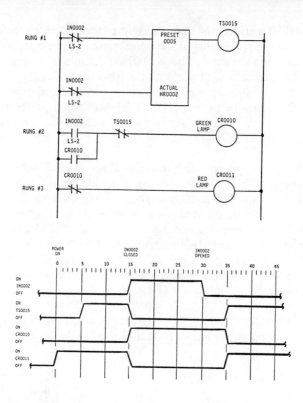

Figure 11-33. Programmed OFF Delay Circuit With N.O.T.O. and N.C.T.C. Contacts

For a retentive timer, two different input devices are connected to the timing and enable circuits as shown in Figure 11-34.

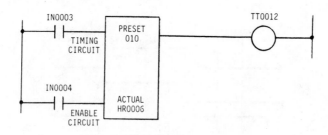

Figure 11-34. Retentive Timer

As long as the enable circuit input device IN0004 is closed, timing can be started by closing input IN0003 in the timing circuit. As long as IN0004 remains closed, IN0003 can be opened and closed with the time being retentive. When the accumulated time equals the preset time, TT0012, timer tenths of seconds, will energize. To reset the timer, IN0004 is opened causing the timer to reset to 0000.

164

SQUARE D COMPANY TIMERS

Figure 11-35 shows a typical Square D Company timer used with their SY/MAX line of PC's.

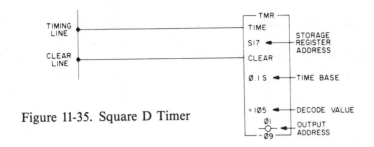

Figure 11-35. Square D Timer

There is a timing line, clear line, and a timer box which contains four pieces of information. The information includes the word address for a storage register that will hold the timer accumulated value, the time base (.01 seconds, 0.1 seconds, and .1 minute), the decode value (preset value) which may be from 0001-9999 (counted using the binary numbering system), and an output address that will be turned on when the accumulated time equals the decode or preset time.

Figure 11-36a shows a standard ON DELAY timer circuit, and Figure 11-36b shows how it would be programmed.

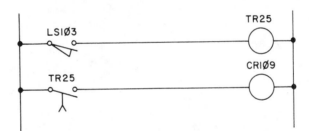

Figure 11-36a. ON Delay Timer Circuit

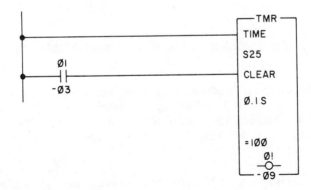

Figure 11-36b. Timer Circuit Programmed Using Square D Format

165

When contact 01-03 closes, the timer will time until the accumulated value in word 25 of the storage register equals the preset time of 100 and turns on output 01-09. If the input device LS-103 is opened before the accumulated value equals the preset or decode value, word 25 in the storage register is cleared to 0000.

To program and duplicate the OFF DELAY timer circuit in Figure 11-37a, an additional rung of logic must be added as shown in Figure 11-37b. This is the same type of circuit that was added for the Westinghouse timer.

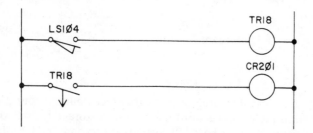

Figure 11-37a. OFF Delay Timer Circuit

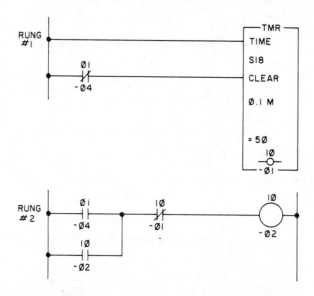

Figure 11-37b. Programmed Circuit to Duplicate OFF Delay Timer

Again a N.C. (EXAMINE OFF) contact is programmed for a N.O. limit switch LS-104. When input 01-04 closes, coil 10-02 will energize and remain energized through its holding contacts until the N.C. contacts 10-01 open. N.C. contacts 10-01 cannot open until input device 01-04 goes open, and the timer times out. The timer is set for .1M (minute) X 50 or 5 minutes.

Additional N.O. contacts addressed 10-02 can be used as timed N.O.T.O. contacts while N.C. contacts addressed 10-02 will serve at timed N.C.T.C.

166

For a retentive or interruptible timer, the circuit in Figure 11-38 is used.

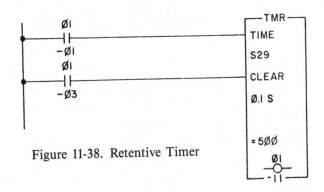

Figure 11-38. Retentive Timer

When contact 01-03 is opened, the timer value will be cleared, S29 = 0000. To start the timer, both contacts 01-01 and 01-03 must be closed. If contact 01-01 is opened and the timer is enabled, the timer will stop timing but will not clear. Reclosing contact 01-01 will allow the timer to continue timing from the value at which it was interrupted, assuming contact 01-03 is still closed. When the value in register S29 is equal to the decode (preset) value of 500, the output address 01-11 will turn ON.

GOULD INC., MODICON DIVISION TIMERS

Figure 11-39 shows the timer format typical for timers programmed for a Modicon 484.

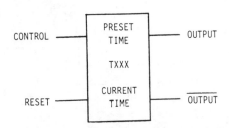

Figure 11-39. Modicon 484 Timer Format

The two lines or nodes to the left are control and reset. The control line controls when the timer times, and the reset line resets the accumulated value of the timer to 000. The timer is enabled when the reset line has power flow and is reset when there is no power flow.

The timer box holds the preset time (001 to 999), time base (T 1.0 = seconds, T 0.1 = tenths of a second, and T .01 = hundredths of a second), and the storage location for the current or accumulated time.

The two right hand nodes or lines are output lines and will provide power to contacts, coils, and so forth. The top line provides power only when the timer accumulated value is equal to the preset value while the bottom line only provides power when the accumulated time is **NOT** equal to the preset time. This output will only stop passing power when the accumulated and preset values are the same.

Figures 11-40a, b, c, and d show how timers would be programmed to duplicate ON and OFF delay timers.

Rather than duplicate the same input device, 1053, a shunt or vertical line can be programmed to tie the control and reset lines together.

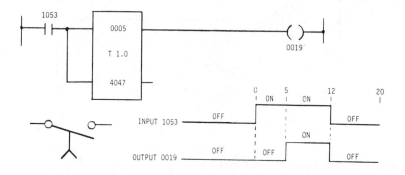

Figure 11-40a. Normally Open Time Closing

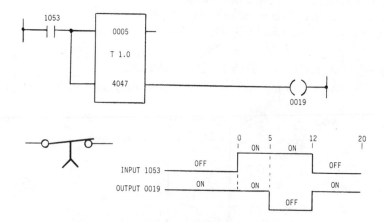

Figure 11-40b. Normally Closed Time Opening

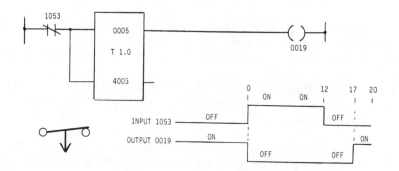

Figure 11-40c. Normally Closed Time Closing

168

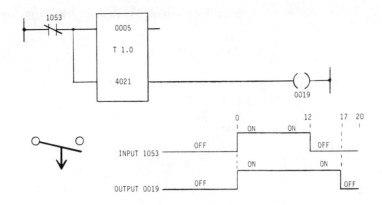

Figure 11-40d. Normally Open Time Opening

NOTE: In Figures 11-40c and d N.C. contacts are programmed for a normally open input device.

For a retentive timer, a different input device is programmed in the control and reset lines as indicated in Figure 11-41.

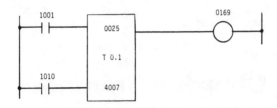

Figure 11-41. Retentive Timer

CASCADING TIMERS

When circuit requirements demand more time than is available from a single timer, 999 or 9999 seconds depending on the PC, two or more timers can be programmed together as shown in Figures 11-42a and b. Programming two or more timers together for more time is called **cascading.**

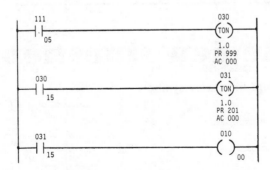

Figure 11-42a. Cascading Timers

169

Total time to turn on 01000 after input 11105 closes is 1,200 seconds, 999 + 201, or 20 minutes.

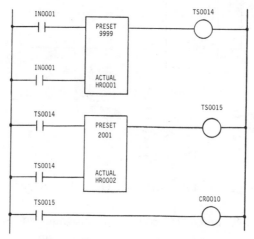

Figure 11-42b. Cascading Timers

Total time to turn on CR0010 after input IN0001 closes is 12,000 seconds, 9999 + 2001, or 200 minutes.

By cascading timers, any required time can be achieved.

Chapter Summary

Though the format is different for different PC's, the basic principles are the same. Preset and accumulated times are stored and compared on each processor scan. When the accumulated value equals the preset value, discrete output devices or internal outputs can be turned ON or OFF. Timers can be programmed for ON DELAY, OFF DELAY, or as RETENTIVE timers. The only limit to the number of timed and instantaneous contacts that can be programmed is memory size. Programmed timers offer a wider range of time setting and greater accuracy than is possible with "HARD WIRED" pneumatic timers.

Review Questions

1. Match the standard time delay symbols.

a. _____

b. _____

c. _____

d. _____

1. ⫟ T.C.

2. ⊣⊢ T.C.

3. ⫸ T.O.

4. ⊣⊢ T.O.

2. The amount of time for which a timer is programmed is called the:
   a. Preset
   b. Set point
   c. Desired time (DT)
   d. All of the above
3. T   F   As scan time increases, so does the accuracy of any programmed timers.
4. When the timing of a device is not reset on a loss of power, the timing is said to be:
   a. Holding
   b. Secured
   c. Retentive
   d. Continuous
5. When more time is needed than can be programmed with one timer, two or more timers can be programmed together. This programming technique is called:
   a. Stacking
   b. Cascading
   c. Doubling
   d. Synchronizing

Chapter 12

Programming Counters

Objectives

After completing this chapter, you should have the knowledge to
- Write a program using up and down counters.
- Define the terms *increment* and *decrement*.

Programmed counters will serve the same function as the mechanical counters that have been used in the past.

Programmed counters can count up, count down, or be combined to count up and down.

Counters are similar to timers, except they do not operate on an internal clock but are dependent on external program sources for counting.

ALLEN-BRADLEY COUNTERS

Like timers, counters compare an accumulated value to a preset value to control circuit functions. Figure 12-1 shows the 16 bit word used by Allen-Bradley for a counter, up or down.

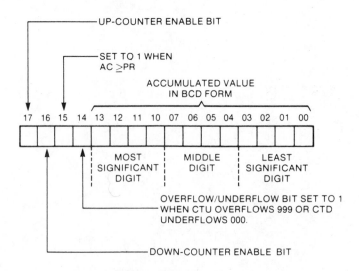

Figure 12-1. Allen-Bradley Counter Format

The first 12 bits, 00-07 and 10-13 octal, are used to store the accumulated value in 3 digit BCD. Unlike a timer that stops timing when the accumulated value equals the preset value, a counter will continue to count up or down, and bit 14 is used as an overflow or underflow status bit and is set to 1 when either condition, overflow (999) or underflow (000), occurs. Bit 15 is set to 1 or ON any time the accumulated value is equal to (=) or greater than (>) the preset value. Bit 16 is the down counter enable bit and is set to 1 any time a down counter rung is true. Bit 17 is the up counter enable bit and is set to 1 or ON when an up counter rung is true.

To activate a counter, up or down, an input device must be used, and the counter only counts up or down when the state of the input device goes from false (open) to true (closed).

Input device 11105 is used in Figure 12-2 to control the count of an up counter. The CTU label stands for count up.

Figure 12-2. Up Counter

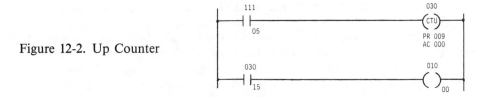

When input 11105 is closed, the counter will **increment** or count UP 1, and the accumulated value (AC) would show 001. The counter cannot increment again until input 11105 is opened and then closed again (transition from false to true). The accumulated value will increment by one with each transition of input 11105. When the accumulated value equals the preset value 009, bit 15 of word 030 will be set to one, and output 01000 will energize. Without a reset device, the counter will continue to count above the preset value of 009 with each false-true transition of input 11105. When the counter accumulated value reaches 999, bit 14 would be set to 1 to indicate an overflow.

To reset a counter, a counter reset (CTR) is added to the circuit as shown in Figure 12-3.

Figure 12-3. Reset Rung

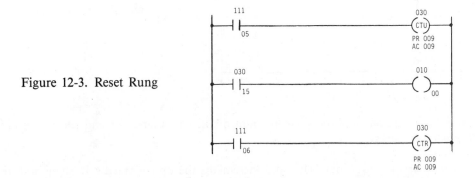

Like the retentive timer reset (RTR), the counter reset (CTR) is given the same word address as the counter it is to reset. When the counter reset rung is true, input 11106 closed, the accumulated value in CTR-030 will be reset to 000, and bit 15 will be cleared to 0 or turned OFF.

Down counter (CTD) programming is illustrated in Figure 12-4.

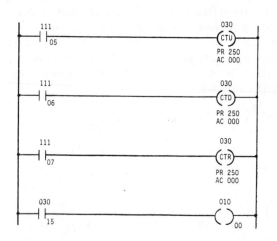

Figure 12-4. Down Counter

With the preset at 100 and the accumulated value preset to 150, the counter will **decrement** or count down from 150 by 1, each time input 11105 goes from false to true (OFF to ON). Since bit 15 is set to 1 or ON any time the accumulated value is equal to (=) or greater than (>) the preset value, output 01000 is ON and will stay on until the counter has counted down to 99. In this circuit, closing input 11106 and activating the counter reset (CTR-040) would reset the accumulated value to 000, not the originally programmed accumulated value of 150. To reset the accumulated value at 150, special data manipulation techniques are used. These techniques will be covered in the next chapter.

Up and down counters can be programmed together as shown in Figure 12-5.

Figure 12-5. Combining Up and Down Counters

A typical application could be to keep count of the cars that enter and leave a parking lot or parking garage.

As a car enters, it trips input 11105 and increments the up counter by 1. Since both the up and down counters as well as the counter reset have the same address, the accumulated value will be the same in all three. When a car leaves the parking lot or garage, it trips input 11106, and the down counter decrements or reduces the accumulated value in the CTU, CTD, and CTR by 1. If the accumulated value reaches the preset value (AC = PR), bit 15 of word 030 is set to 1, and output 01000 is energized. Output 01000 could be a "LOT

174

FULL" sign or any other output device that may be used to indicate all parking spaces are full.

Up, down, and up-down counters are retentive (hold their count) on power failure or if programmed after an MCR.

WESTINGHOUSE COUNTERS

Figure 12-6 is a typical Westinghouse up counter.

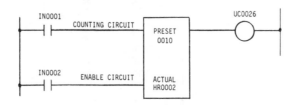

Figure 12-6. Westinghouse Up Counter Format

The counting circuit will increment, increase by one, the actual count each time the counting circuit goes from non-conducting to conducting or from OFF to ON. Of course, like timers, the counting circuit can only increment the counter if the enable circuit is ON. With input IN0002 in the enable circuit ON, the counter will start from 0000 and increment the count by one each time input device IN0001 in the counting circuit goes from false (OFF) to true (ON). When the actual (accumulated) count stored in HR0002 equals the preset count, output UC0026 will be energized. Any contacts labeled UC0026 in other rungs of the circuit would open and/or close when UC0026 energized. The counter is reset to 0000 when the enable circuit is opened and is held at 0000 until the enable circuit is again closed or true.

Figure 12-7 shows a typical Westinghouse down counter.

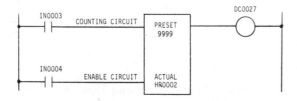

Figure 12-7. Down Counter

Down counters count from a preset value down to 0000. When the actual count reaches 0000, output DC0027 is energized. When input IN0004 is closed, the counter is enabled and will count down, **decrement,** from 9999 or other preset value down to 0000 each time the input device IN0003 in the counting circuit goes from false (OFF) to true (ON). When the enable circuit is opened, the accumulated value in HR0002 is reset and held at the preset value instead of 0000 like an up counter. When the enable circuit is energized, each false to true transition of the counting circuit will count down from the preset value to 0000. When the actual (accumulated) count reaches 0000, output DC0027 is energized.

Up and down counters can be programmed together as shown in Figures 12-8a and b to count products as they enter a conveyor line (count up) and as they leave the line (count down).

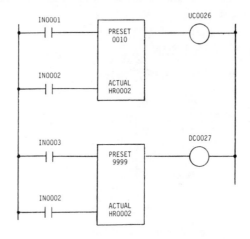

Figure 12-8a. Combining Up and Down Counters

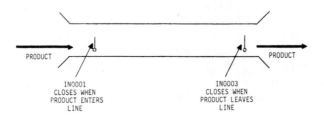

Figure 12-8b. Applying Up and Down Counters

Notice that the same holding register (storage word) HR0002 is used for both up counter UC0026 and down counter DC0027.

By sharing the same holding register, the actual value stored in HR0002 of the up counter, which is programmed first, determines the actual value in HR0002 of the down counter, regardless of the preset value of DC0027.

As a product enters the conveyor line, input IN0001 is activated, false to true, and the actual count in HR0002 increments from 0000 to 0001. The next product would increment HR0002 to 2, and so on. If 8 products had entered the line before any were removed, the actual value in HR0002 would be 0008. The first product to leave the line and activate IN0003 would decrement the actual count in HR0002 from 0008 to 0007. The next product that left the line and activated IN0003, false to true, would again decrease the accumulated count in HR0002 from 0007 to 0006.

The indicator lamps, outputs DC0027 (green) and UC0026 (red) in Figure 12-8b indicate an empty line when DC0027 has counted down to 0000 or a full line when the preset value 0010 of UC0026 is reached. Input device IN0002 is used as a start/reset button.

Up and down counters are retentive on power failure.

SQUARE D COMPANY COUNTERS

Figure 12-9 shows the unique approach the Square D Company uses for programming counters with their SY/MAX 100 and 300 Series PC's.

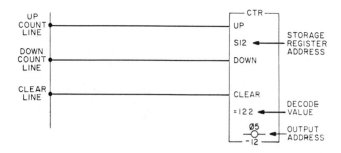

Figure 12-9. Square D Counter Format

The counter is a combination up and down counter. Using only the up count and clear lines, the counter is an up counter only. Using the down count and clear lines, it can be a down counter only. By using all three lines, it is a combination up and down counter.

Figure 12-10 illustrates how the counter would be programmed as an up counter only.

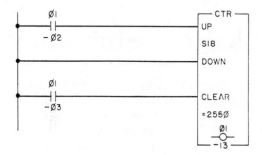

Figure 12-10. Up Counter Only

177

If the clear circuit input device 01-03 is closed, activating input device 01-02 in the up line, transition from OFF to ON will cause the accumulated value in storage word 18 to increment by one, from 0000 to 0001. Each subsequent transition of input device 01-02 would increase the value in storage register 18 until the accumulated value equaled the preset value of 2550, at which time output 01-13 would be energized.

To program a down counter, special data manipulation techniques are used. Data manipulation is covered in the next chapter, and an example of programming a down counter will be used.

Figure 12-11 illustrates an up-down counter by using the up, down, and clear lines.

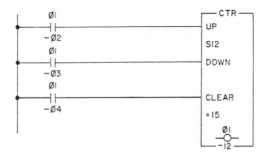

Figure 12-11. Up and Down Counter

With input device 01-04 in the clear line open, the value in storage word 12 is cleared to 0000, and output 01-12 is de-energized. When input device 01-04 in the clear line is closed, the counter is activated. Opening and closing input device 01-02 in the up line will increment the value stored in storage register 12 by 1. Open and closed transitions of input device 01-03 in the down line will decrement (reduce) the value in S12 by 1. Output 01-12 will not energize until the accumulated value in S12 equals the decode (preset) value.

Up, down, and up-down counters will hold the accumulated value on a power failure or when programmed after an MCR (master control relay).

Chapter Summary

Programmed counters give added flexibility and control for electrical process equipment and/or driven machinery. Like timers, counters store values in Binary or Binary Coded Decimal (BCD) format for the preset and accumulated count. For up counters the processor compares the preset and accumulated values on each scan and updates the I/O section on the scan that the accumulated value equals the preset value. For down counters the processor updates the I/O section when the accumulated value is 0 for some PC's and when the accumulated value is equal to or greater than the preset with other PC's. Counters count, increment or decrement, when the count rung transition is from false to true.

Review Questions

1. In Figure 12-A, switch 11105 is now open. When switch 11105 is closed, counter 030 will:

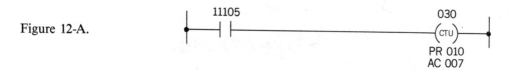

Figure 12-A.

    a. Increment by 1
    b. Decrement by 1
    c. Not count and the accumulated value will remain at 7

2. The reset rung shown in Figure 12-B will reset counter 30:

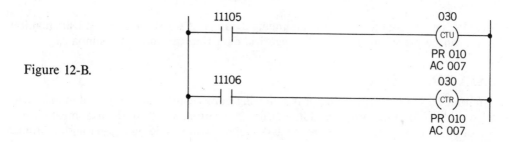

Figure 12-B.

    a. Automatically when the count reaches 010
    b. Automatically when the count reaches 011
    c. Only when the count reaches 999
    d. Only when switch 11106 is closed
    e. Only when switch 11106 is closed then opened

3. T   F   Counters can count past their preset value.

4. Output 02000 as shown in rung 2 of Figure 12-C, will be true:

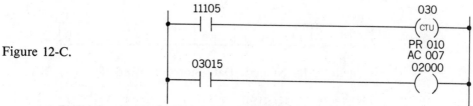

Figure 12-C.

    a. Only when the count is equal to the preset value
    b. When the count is equal to or greater than the preset value
    c. When the count is less than the preset value
    d. When the accumulated value reaches 999 and overflows
    e. When the count goes to 011

5. T   F   Up and down counters can be programmed together to count up and down.

Chapter 13

Data Manipulation

Objectives

After completing this chapter, you should have the knowledge to
- Explain what data transfer is.
- Define the term *writing over*.
- Write a rung of logic that transfers data from one word to another.
- Identify the standard data compare instructions.
- Write logic that compares data to control an output.

Most PC's now have the ability to manipulate data that is stored in memory. Data manipulation can be placed in two broad categories: data transfer and data compare.

DATA TRANSFER

Data transfer consists of moving or transferring numeric information stored in one memory word location to another word in a different location. Words in the user memory portion of the processor may be referred to as data table words, holding register and/or storage register words depending on the PC.

Figures 13-1a and b illustrate the concept of moving numerical data from one word location to another word location. Figure 13-1a shows that numeric (binary) data is stored in word 18 and that no information is currently stored in word 31.

	16	15	14	13	12	11	10	9	8	7	6	5	4	3	2	1
WORD 18	1	1	0	1	0	0	1	1	0	1	0	1	1	1	0	1

	16	15	14	13	12	11	10	9	8	7	6	5	4	3	2	1
WORD 31	0	0	0	0	0	0	0	0	0	0	0	0	0	0	0	0

Figure 13-1a. Numeric Data Stored in Word 18

	16	15	14	13	12	11	10	9	8	7	6	5	4	3	2	1
WORD 18	1	1	0	1	0	0	1	1	0	1	0	1	1	1	0	1

	16	15	14	13	12	11	10	9	8	7	6	5	4	3	2	1
WORD 31	1	1	0	1	0	0	1	1	0	1	0	1	1	1	0	1

Figure 13-1b. Data Transferred From Word 18 Into Word 31

After data transfer (Figure 13-1b), word 31 now holds the exact or duplicate information that is in word 18. If word 31 had information already stored, rather than all 0's, the information would have been replaced. When new data replaces existing data in a word after a transfer, it is referred to as **writing over** the existing data.

Chapter 12, programming counters, discussed Allen-Bradley down counters and indicated that by using special programming techniques, the counter accumulated value could be reset to some value other than 000. This is accomplished by using two instruction keys on the industrial terminal (programmer): **GET** [G] and **PUT** (Put).

GET instructions tell the processor to go get a value stored in some word. The GET instruction shown in Figure 13-2a would tell the processor to get the value 150 that is stored in word 020.

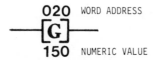

Figure 13-2a. Allen-Bradley GET Instruction

The PUT instruction tells the processor where to put the information it obtained from the GET instruction.

The GET-PUT instructions in Figure 13-2b would tell the processor to get the numeric value 150 stored in word 020 and put it into word 070. PUT instructions must be preceded by a GET instruction.

Figure 13-2b. GET-PUT Instruction

Figure 13-3 shows an Allen-Bradley down counter circuit with a rung containing GET [G] and PUT (Put) instructions for resetting the down counter to an accumulated value of 150.

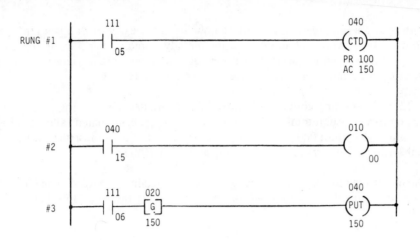

Figure 13-3. Resetting a Down Counter Using a GET-PUT Instruction

After programming input device 11106, the GET key [G] is pressed, and the GET [G] instruction can now be addressed with an unused word in memory. In this example, word 20 is used. Next, the desired numeric value of 150 is entered and will be stored in word 20. The PUT key (Put) is now pressed, and the address 040 is entered because that is where the accumulated value for down counter 040 is stored.

NOTE: Remember that the accumulated values are always stored in the word address of the Allen-Bradley timer or counter while the preset values are automatically stored 100 words higher.

The value entered in word 020 (150) automatically is placed below the PUT symbol on the programmer CRT.

Rung 3 now says: If input device 11106 is closed, **GET** the numeric value stored in word 020 (150) and **PUT** it into word 040.

NOTE: Remember with Allen-Bradley down counters, bit 15 is set to 1 or ON any time the accumulated value is greater than (>) or equal to (=) the preset value. Output 01000 in Figure 13-3 would be ON until the time counted down to 099 (less than and not equal to 100).

Now, when the counter has counted down from 150 to 099 and de-energized output 01000, the counter accumulated value can be reset to 150, rather than 000, by closing input 11106.

With the GET-PUT instructions, data in any storage word can be transferred to other words for accumulated **or** preset values. An example might be a counter that counts boards at a sawmill. Let's say we are stacking 2 X 4's and want 400 in the stack before we bind and move the stack. When the mill is producing 2 X 6's, however, we only want 250 boards in a stack. Figures 13-4a and b show how to change the preset value of an up counter for each different lumber size by using pushbuttons.

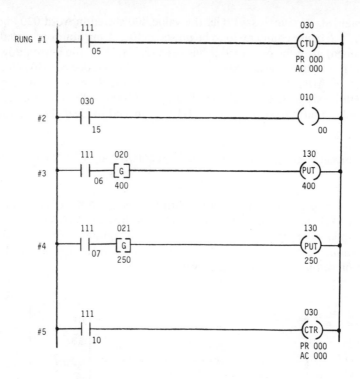

Figure 13-4a. Changing Preset Value With GET-PUT Instructions

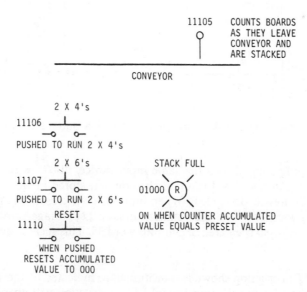

Figure 13-4b. Pushbuttons Used to Change Preset Values

Up counter 030 is initially programmed with no preset, 000. The preset will be determined by whichever pushbutton is depressed, 2 X 4's — 11106 or 2 X 6's — 11107. If 2 X 4's are to be run, pushbutton 11106 is pushed, enabling the GET-PUT statement of rung 3. When

183

this rung is enabled, or true, it says: Get the value 400 stored in word 020 and put it into word 130. This will cause counter 030 to be preset at 400. A pushbutton is used so the rung will go false (open) after the preset value has been set. Holding the button down and keeping rung 3 true holds the value at 400, and transitions of input device 11105 could not increment the counter. After 400 boards have been counted (PR = AC), bit 15 or word 030, up counter, will be set to 1, and the stack full light, 01000, will come on. After the stack has been moved, counter reset pushbutton 11110 is pushed to clear the accumulated value to 000.

To change the preset value of up counter 030 from 400 to 250, the 2 X 6's pushbutton 11107 is depressed.

Square D Company uses a **LET** instruction with their SY/MAX PC's to achieve the same results. With a LET instruction, a storage word is addressed, and then the desired value is entered. Chapter 12 skipped over the Square D down counter and indicated that it took a special programming technique. Figure 13-5 shows a down counter and a LET rung programmed for presetting the accumulated value.

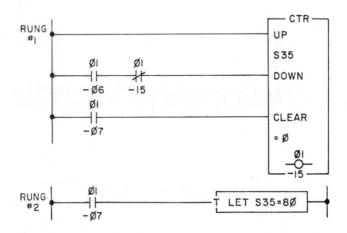

Figure 13-5. Presetting the Accumulated Value of a Square D Down Counter

The LET rung is programmed with the same input device, 01-07, as the clear line. With input 01-07 open, the accumulated value of the counter, stored in word 35, is cleared to 0000, and output address 01-15 is OFF. When input 01-07 is closed, the timer is enabled, and the LET rung goes true. The LET statement says: Let storage word 35 = 80. With the rung true, a value of 80 is entered or put into word 35. This sets the accumulated value of the down counter at 80.

NOTE: The LET instruction shown is transitional, as indicated by the T in the left margin of the LET box. A transitional LET instruction will operate only once on each transition (open-close) of input device 01-07. This allows the accumulated value to be set, but not held, at 80.

Each time 01-06 has a transition from false to true, the counter will decrement by 1. After 80 transitions of input 01-06, the accumulated value will equal zero (AC = PR), and output 01-15 will turn ON. When output 01-15 energizes, it opens the normally closed contact

184

in the down line disabling the counter. Any additional transition of contacts 01-06 will be ignored. To turn OFF output 01-15 and reset the counter, input device 01-07 is opened and then closed again. With the accumulated value again set to 80 by the LET rung, input device 01-06 will again decrement the counter by 1 with each transition from OFF to ON or FALSE to TRUE.

A regular non-transitional LET instruction is shown in Figure 13-6.

Figure 13-6. Non-Transitional LET Statement

When contacts 01-04 and 01-08 are closed, the value 1535 will be preset into storage register S42. The LET will be performed with each scan of the processor memory as long as 01-04 and 01-08 remain closed.

Data transfer from one storage word to another is illustrated in Figure 13-7.

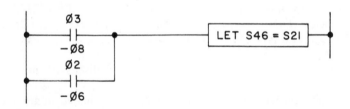

Figure 13-7. Transferring Data From One Storage Word to Another

Closing either contact 03-08 or 02-06, the value in storage register S21 will be transferred into storage register S46.

Westinghouse uses an **MV** instruction to move a value stored in one register word into another register word. Figure 13-8 illustrates an MV instruction typical of the Westinghouse 700/900 Series PC's.

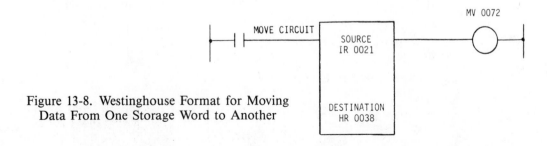

Figure 13-8. Westinghouse Format for Moving
Data From One Storage Word to Another

185

The source, or where the data to be copied is stored, may be a holding register (HR), input register (IR), output register (OR), and so on. The destination, or where the data is to be moved, can be a holding register (HR), output register (OR), or an output group (OG).

Again, the principle is the same. When the move circuit is true, the data (numeric information) stored in the source IR0021 will be copied into the destination address HR0038.

It does not matter whether it is a GET-PUT, LET, MV instruction, or any other designated instruction by the different PC manufacturers. They are all DATA TRANSFER instructions, and the object is to move numerical information from one word into another. To illustrate the concept, the data transfer for entering accumulated and preset values into counters was used. This is by no means the only application for a data transfer instruction. While using a given PC and becoming more familiar with its operation and capabilities, data transfer will become a powerful instruction that will have many applications.

DATA COMPARE

Data compare opens a whole new realm of programming possibilities and demonstrates why PC's will replace most, if not all, "hard wired" systems in the very near future.

Data compare instructions will, as the name implies, compare the data stored in two or more words and make decisions based on the program instructions.

Numeric values in two words of memory can be compared for **LESS THAN ($<$), EQUAL TO ($=$), GREATER THAN ($>$), LESS THAN or EQUAL TO ($\leq$), GREATER THAN or EQUAL TO ($\geq$),** and **NOT EQUAL TO ($\neq$)** depending on the PC.

Data compare concepts have already been used when we discussed timers and counters.

The ON DELAY timer turned ON an output when the accumulated value equaled the preset value (AC $=$ PR). What actually happened to turn ON an output was that the accumulated numeric data in one memory word was compared to the preset value in another word on each scan of the processor. When the processor saw that the accumulated value equaled the preset value (AC $=$ PR), it turned ON the output.

Additional programming instructions can compare memory words and turn on outputs when the values are less than ($<$), equal to ($=$), greater than ($>$), and so on.

To graphically demonstrate how data compare instructions can be used, consider the "hard wired" circuit in Figure 13-9. This circuit uses 3 pneumatic time delay relays to start up a conveyor system in inverse order 4-3-2-1.

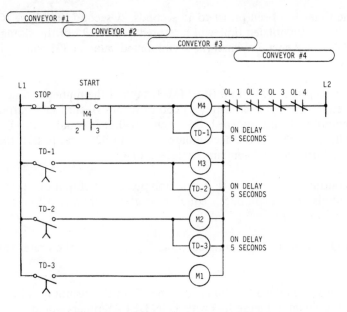

Figure 13-9. Hard Wired Conveyor System

The same circuit can be programmed on a PC using only one internal timer and two data compare statements as shown in Figure 13-10. The Square D format will be used as an example.

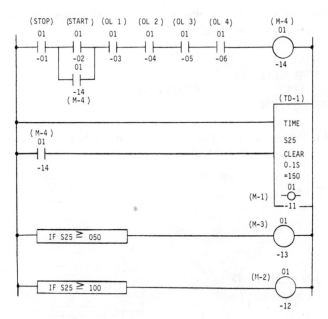

Figure 13-10. Square D Data Compare Format

Assuming that the stop button 01-01 and the overloads 01-03, 01-04, 01-05, and 01-06 are closed, when the start button (01-02) is pushed, M-4 contacts 01-14 will energize and holding contacts 01-14 will close and hold the circuit in. M-4 contacts 01-14 close and enable

187

the timer. The timer has been preset to 15 seconds, .1 second time base X preset 150 = 15.0 seconds. The accumulated time will be stored in word 25 of the storage register, and output 01-11 (M-1) will energize when the accumulated value in S25 equals the preset value of 15 seconds.

The IF instruction preceding output 01-03 (M-3) says: If the numeric value stored in word S25 becomes equal to or greater than (≥) 050 (5 seconds), turn ON output 01-13 (M-3). When the accumulated value in word S25 reaches 050, it is equal to (=) 050, and output 01-13 (M-3) will turn ON. As the accumulated value in S25 goes to 051, the value is now greater than (>) 050, so M-3 remains energized or ON.

As the timer continues to time, the IF instruction preceding output 01-12 (M-2) will be true when the accumulated value in S25 is equal to or greater than (≥) 100 (10 seconds), and output 01-12 will turn ON.

M-1 will turn ON when the accumulated value in S25 equals the preset value of 150 (15 seconds).

Not only does this give added flexibility for timers, but this technique also saves memory. Programming the circuit in Figure 13-9 with 3 ON DELAY timers would have used 18 words of user memory, as each timer requires 6 words of memory. IF and LET statements only use 3 words each, so by programming the circuit in Figure 13-10 only 12 memory words are used for the timing function. Six words are required for the timer and 3 words for each of the two IF statements (6 + 6 = 12).

The data comparisons symbols that are used with an IF instruction with the Square D SY/MAX 100 and 300 PC's are as follows:

> = EQUAL TO
> ≠ NOT EQUAL TO
> ≥ GREATER THAN OR EQUAL TO
> < LESS THAN.

Figure 13-11a is a circuit that illustrates how each instruction operates.

Figure 13-11a. Square D
Data Comparisons

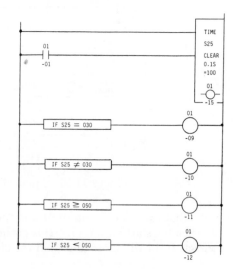

The timer is set for 10 seconds, .1 seconds X 100.

When power is applied, but before 01-01 is closed to activate the timer, outputs 01-10 and 01-12 will be energized. The IF statement preceding output 01-10 is TRUE if storage word S25 is not equal ($\neq$) to 030. With 01-01 OPEN, S25 is reset to 000 and is NOT equal to 030. Output 01-12 is energized because the IF statement is TRUE any time the value in S25 is less than (<) 050.

When input 01-01 closes, the timer is enabled and starts to time. At time 030, the IF statement preceding 01-09 goes TRUE since the accumulated value in S25 is equal to (=) 030, and output 01-09 is energized. This will only be true while the accumulated value in S25 is at 030. When it advances to 031, the IF statement goes FALSE and 01-09 goes OFF. Output 01-10 which was ON because of the not equal to ($\neq$) IF statement will go OFF for 1 second because the IF statement was false when the value in S25 was equal to (=) 030.

At 050, 5 seconds, output 01-11 will turn ON since the IF statement preceding it goes true when S25 is equal to or greater than ($\geq$) 050. The rung is true when the accumulated value in S25 is equal to (=) 050 and remains true as long as the accumulated value is 050 or greater. Output 01-11 will remain ON until the timer is cleared and the accumulated value in S25 is reset to 000.

Output 01-12 that has been ON goes OFF at 050 because the IF statement that precedes it is only TRUE when the value of S25 is less than (<) 050.

Output 01-15, the timer output, will come on at 100 when the accumulated value equals the preset value. The time chart in Figure 13-11b illustrates the ON and OFF states of the outputs in relation to time and the IF instructions.

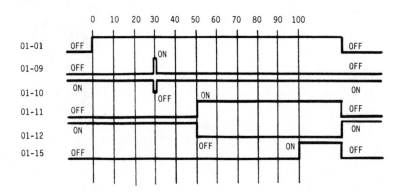

Figure 13-11b. Time Chart for Data Comparisons

Data comparisons can also be made with the value in one storage register to a value in another storage register (Figure 13-12).

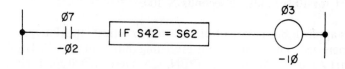

Figure 13-12. Comparing Data in Storage Registers

If input device 07-02 is CLOSED and IF the value in storage word 42 (S42) is EQUAL TO (=) the value in storage word 62 (S62), output 03-10 will energize.

In the timer circuit (Figure 13-10), a number in a storage register was compared to a constant. Figure 13-13 is another example of the comparison of a storage register value to a constant. A value in a storage register can also be compared to the results of a math operation. Math (arithmetic) operations will be covered in Chapter 14.

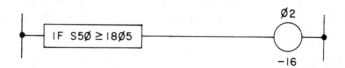

Figure 13-13. Comparing Data in a Storage Register to a Constant

Data comparison instructions can be programmed in series, parallel, or series parallel on the same rung for added circuit control. Figures 13-14a and b show IF statements programmed in series and parallel.

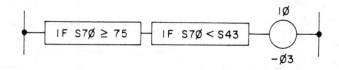

Figure 13-14a. Data Comparisons Programmed in Series

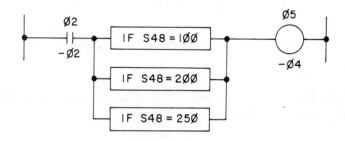

Figure 13-14b. Data Comparisons Programmed in Parallel

190

The basic Allen-Bradley data comparison instructions on the programmer are less than (<) and equal to (=). But by combining instructions, data comparisons can also be made for greater than (>), less than or equal to (≤), greater than or equal to (≥), and not equal to (≠).

A GET [G] instruction must be programmed preceding any data comparison instruction.

An equal to (=) data comparison example is shown in Figure 13-15.

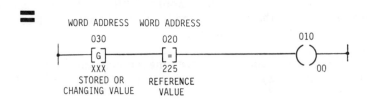

Figure 13-15. Allen-Bradley EQUAL TO Comparison

First a GET instruction, [G], is programmed and given the address of the word that stores the value to be compared. The value, represented by XXX, could be a stored value in memory or a changing accumulated value of a timer or counter. Next, the data comparison instruction is entered and given the address of the word in which the reference or comparison value is to be stored. In the example (Figure 13-15), output 01000 will turn ON when the value in word 030 is equal to the reference value 225 in word 020. Output 01000 will de-energize if the value in 030 goes to 226 since the value would no longer be equal to the reference value of 225.

For a less than data comparison (<), the circuit is programmed as shown in Figure 13-16.

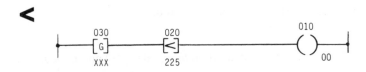

Figure 13-16. Allen-Bradley LESS THAN Comparison

Output 01000 will be ON as long as the value of word 030 is less than (<) the reference value 225 of word 020 and will go OFF when the value is equal to or greater than 225.

To program a greater than data comparison, a sort of reverse logic is used. As stated earlier, there are only two basic data comparison instructions, [=] and [<], on the Allen-Bradley programmer. The less than [<] instruction is used for a greater than comparison and addressed with the word that is to be compared. The GET [G] instruction will be addressed with the word that will store the reference value. A greater than circuit is shown in Figure 13-17.

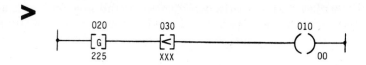

Figure 13-17. Allen-Bradley GREATER THAN Comparison

For a greater than comparison, the value in word 030 (the less value) is compared to the reference value of word 020 (the get value) and the rung will only be TRUE when the less value is greater than the GET value of 225.

For a less than or equal to (≤) comparison, the rung is programmed with one GET instruction followed by less than (<) and equal to (=) instructions in parallel (Figure 13-18).

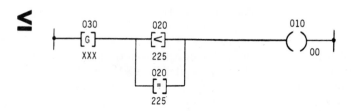

Figure 13-18. Allen-Bradley LESS THAN OR EQUAL TO Instruction

Output 01000 will be ON as long as the value in word 030 is less than (<) 225. When the value in word 030 reaches 225, the (=) instruction is true but will go false at 226.

NOTE: Both the (<) and (=) instructions can use the same word address and reference number.

Figures 13-19 and 13-20 illustrate greater than or equal to (≥) and not equal to (≠) circuits.

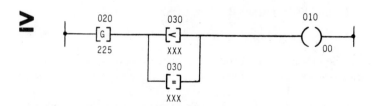

Figure 13-19. Allen-Bradley GREATER THAN or EQUAL TO Instructions

Output 01000 comes on when the value in word 030 equals the reference value 225 in word 020 and will stay on if the value goes to 226 since this is now greater than 225.

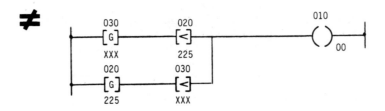

Figure 13-20. Allen-Bradley NOT EQUAL TO Instruction

Greater than (>) and less than (<) statements are programmed in parallel so that output 01000 will energize as long as the value in 030 is less than (<) 225 and will de-energize when the value in 030 is equal to the reference value (225) in word 020. If the value in 030 goes to 226, it is now greater than 225, and output 01000 will again be energized.

Westinghouse 700/900 PC's also use only two instructions for data comparisons. They are the equal to **(EQ)** and the greater than or equal to **(GE)** instructions.

Figure 13-21 shows the format for the EQ and GE instructions.

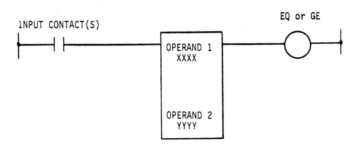

Figure 13-21. Westinghouse Data Compare Format

NOTE: An operand is either of two numbers used in a basic computation to produce an answer. For example: 2 X 3 = 6, 2 and 3 are the operands. Numeric data required for the operation of a special PC function can also be referred to as an operand.

Operand 1 is the number to be compared. The value can be a holding register (HR), input register (IR), or an output register (OR). An example is as follows: A = B, A is operand 1.

Operand 2 is the number to which operand 1 (A) is compared. For example, if A = B, B is operand 2. Operand 2 can be a constant value or a value in a holding register (HR), input register (IR), or an output register (OR).

The EQUAL TO comparison (EQ) instruction works as follows: When the enable circuit conducts and operand 1 equals operand 2, the EQ coil energizes. If operand 1 and 2 are not equal or if the enable circuit does not conduct, the coil is de-energized.

By programming contacts from the EQ coil, equal to (A = B) and not equal (A ≠ B) comparisons can be made (Figure 13-22).

FUNCTION	EQUATION	CIRCUIT CONDUCTS WHEN EQUATION IS TRUE
EQUAL (EQ)	A = B	EQ
NOT EQUAL	A ≠ B	EQ

Figure 13-22. Westinghouse EQUAL TO and NOT EQUAL TO Comparisons

The GREATER THAN or EQUAL TO comparison (GE) instruction functions as follows: When the enable circuit conducts and operand 1 is greater than or equal to operand 2, GE coil energizes. If operand 1 is less than operand 2 or if the enable circuit does not conduct, coil GE is de-energized.

Programming contacts from a GE coil (greater than or equal to) allow for data comparisons of GREATER THAN or EQUAL TO (A ≥ B) and LESS THAN (A < B) as shown in Figure 13-23.

FUNCTION	EQUATION	CIRCUIT CONDUCTS WHEN EQUATION IS TRUE
GREATER THAN OR EQUAL TO (GE)	A ≥ B	GE
LESS THAN	A < B	GE

Figure 13-23. Westinghouse GREATER THAN OR EQUAL TO and LESS THAN Comparison

By combining EQ and GE instructions, and contacts from each, data comparisons for GREATER THAN (>) and LESS THAN or EQUAL TO (≤) are possible (Figure 13-24).

FUNCTION	EQUATION	CIRCUIT CONDUCTS WHEN EQUATION IS TRUE
GREATER THAN	$A > B$	 GE EQ —| |—————|/|—
LESS THAN OR EQUAL TO	$A \leqslant B$	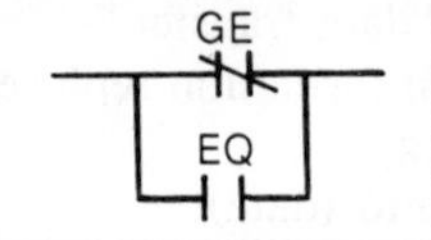 GE —|/|— EQ —| |—

Figure 13-24. Westinghouse GREATER THAN and
LESS THAN OR EQUAL TO Instructions

Like the data transfer feature, once the operator becomes familiar with a specific PC, many applications and advantages to the data compare instructions will emerge.

Chapter Summary

Even though the format and/or instructions varied with each of the three PC manufacturers used as examples, the concepts of data manipulation remained the same. Data manipulation enables the operator to transfer data from one word location to another while data comparison allows the value in one word to be compared to another word or a constant value.

Both data transfer and data comparison instructions give new dimension and flexibility to motor control circuits, and the application of either is only limited by operator imagination.

Review Questions

1. Define the term "data transfer."
2. When numerical information replaces data that already existed in a memory location is it referred to as:
   a. Exchanging info (data)
   b. Replacement programming
   c. Blanket move
   d. Writing over
3. Match the symbols and their correct definitions.
   a. > _____   1. Less than 1
   b. < _____   2. Less than
   c. = _____   3. Less than or equal to
   d. ≥ _____   4. Greater than
   e. ≠ _____   5. Greater than or equal to
   f. ≤ _____   6. Equal to
                       7. Not equal to
                       8. Not equal to 1
                       9. Greater than 1
4. Define the term "data compare."
5. Give an example of how the data compare instruction could be used.

Chapter 14
Arithmetic

Objectives

After completing this chapter, you should have the knowledge to
- Express negative numbers in 2's complement.
- Add signed numbers.
- Convert a negative binary display to its decimal equivalent.
- Complete a subtraction problem using 2's complement and addition.
- List the four standard math functions available with most PC's.
- Discuss the math functions and give examples of how they could be used.

2's COMPLEMENT

Virtually all programmable controllers, computers, and other electronic calculating equipment perform counting functions using the binary system. For those PC's that can be programmed to perform arithmetic functions, a method of representing both positive (+) and negative (−) numbers must be used. The most common method is 2's complement. Two's complement is simply a convention for binary representation of negative decimal numbers.

Before going any further with a discussion of 2's complement, a review of adding binary numbers will be helpful.

In decimal addition, numbers are added according to an addition table. A partial addition table is shown in Figure 14-1.

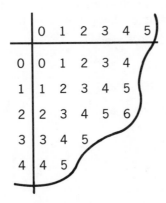

	0	1	2	3	4	5
0	0	1	2	3	4	
1	1	2	3	4	5	
2	2	3	4	5	6	
3	3	4	5			
4	4	5				

Figure 14-1. Decimal Addition System

To use the table, the first number to be added is located on the vertical, the second number on the horizontal and the sum, or total, is where two imaginary lines would intersect. For example, 3 + 2 = 5 as shown in Figure 14-2.

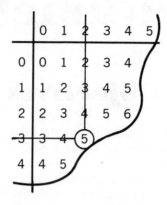

Figure 14-2. Adding 2 and 3

For binary addition, a similar addition table could be constructed. This table will be quite small because the binary system only has two digits, 1 and 0 (Figure 14-3).

Figure 14-3. Binary Addition Table

	0	1
0	0	1
1	1	1 0

Again, to use the table, the first number (digit) to be added is located on the vertical, the second digit on the horizontal, and the sum, or total, is where two imaginary lines intersect. Figure 14-4 shows an example adding 1 + 0 = 1.

Figure 14-4. Adding Binary 1 and 0

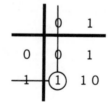

198

Notice that if 1 and 1 are added, the table shows 1 0 not 2 as might be expected. 1 0 is the binary representation of 2 (Figure 14-5).

Figure 14-5. Binary Representation of 2

Figure 14-6 shows how binary numbers 1011_2 and 110_2 would be added.

Figure 14-6. Adding Binary Numbers

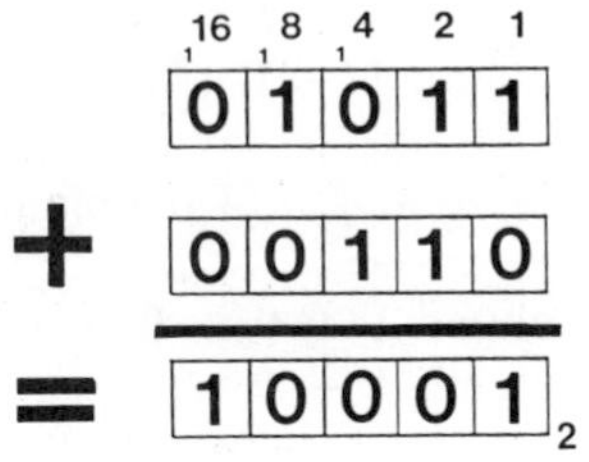

In the 1's Column $1 + 0 = 1$
In the 2's Column $1 + 1 = 0$ with a carry of 1
In the 4's Column $1 + 0 + 1 = 0$ with a carry of 1
In the 8's Column $1 + 1 + 0 = 0$ with a carry of 1
In the 16's Column $1 + 0 + 0 = 1$

The sum (total) of 1011_2 and 110_2 then is 1001_2

To verify our results we can convert the binary numbers to decimal equivalent numbers and add as shown in Figure 14-7.

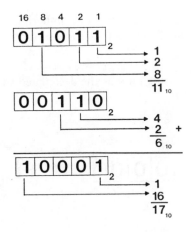

Figure 14-7. Converting Binary Numbers to Decimal Equivalents

199

Another example of adding binary numbers is shown in Figure 14-8 where 11011_2 and 11_2 are added.

$$\begin{array}{r} {\scriptstyle 1\ 1} \\ 11011 \\ +\quad 11 \\ \hline 11110 \,_2 \end{array}$$

Figure 14-8. Addition of Binary Numbers

In the 1's Column $1 + 1 = 0$ with a carry of 1
In the 2's Column $1 + 1 + 1 = 1$ with a carry of 1

NOTE: $1 + 1 + 1 = 3$ the binary equivalent of 3_{10} is 11_2

In the 4's Column $1 + 0 + 0 = 1$
In the 8's Column $1 + 0 = 1$
In the 16's Column $1 + 0 = 1$

The sum of 11011_2 and 11_2 is 11110_2.

To verify, convert the binary numbers to decimal numbers and add as shown in Figure 14-9.

Figure 14-9. Comparing Binary and Decimal Addition

$$\begin{array}{rcl} 11011_2 & = & 27_{10} \\ +\quad 11_2 & = & 3_{10} \\ \hline 11110_2 & = & 30_{10} \end{array}$$

To represent negative numbers using the binary numbering system, one bit is designated as a signed bit. If the designated bit is a 0 (zero), the number is positive. If the bit is a 1, the number is negative.

Using a 4 bit word length and using bit 4 as the designated signed bit, 0001_2 would represent +1 decimal (Figure 14-10).

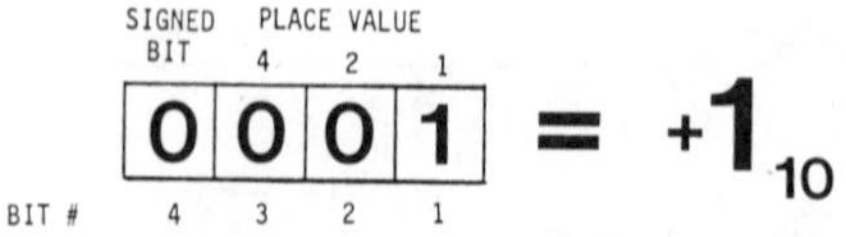

Figure 14-10. Four Bit Word With Signed Bit

With a 4 bit word length and a bit 4 designated as a signed bit, the largest positive number that could be displayed would be +7 as shown in Figure 14-11.

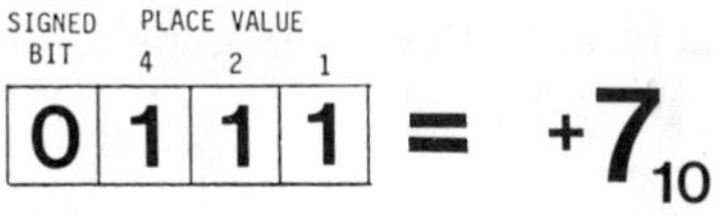

Figure 14-11. Signed Binary Representation of Decimal Number +7

SIGNED BIT   PLACE VALUE   4   2   1

$$\boxed{0\;|\;1\;|\;1\;|\;1} = +\mathbf{7}_{10}$$

An 8 bit word could represent a positive +127 (Figure 14-12) and a 16 bit word would be +32,767.

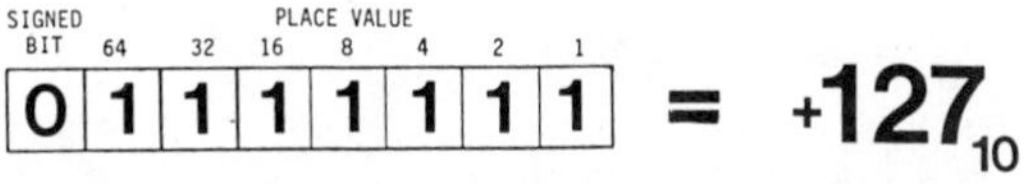

SIGNED BIT   64   32   PLACE VALUE   16   8   4   2   1

$$\boxed{0\;|\;1\;|\;1\;|\;1\;|\;1\;|\;1\;|\;1} = +\mathbf{127}_{10}$$

Figure 14-12. Eight Bit Word With a Signed Bit

To display a negative binary number requires that the same value positive number be complemented (all 1's changed to 0's and all 0's changed to 1's) and a value of 1 added. The result is 2's complement of the number.

Figure 14-13 shows the steps to express −5 in 2's complement using a 4 bit word.

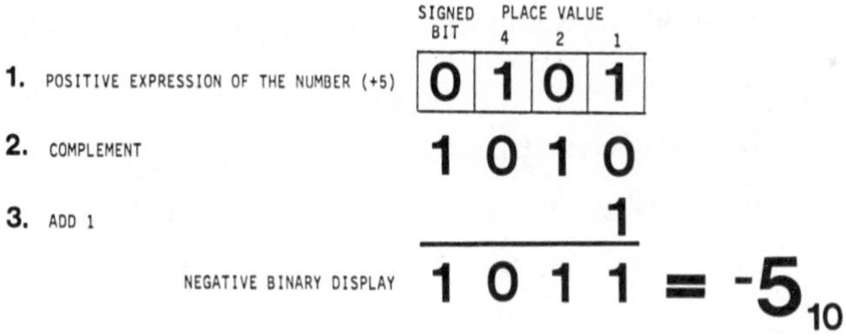

SIGNED BIT   PLACE VALUE   4   2   1

1. POSITIVE EXPRESSION OF THE NUMBER (+5) $\boxed{0\;|\;1\;|\;0\;|\;1}$

2. COMPLEMENT   1 0 1 0

3. ADD 1   1

NEGATIVE BINARY DISPLAY $1\;0\;1\;1 = -\mathbf{5}_{10}$

Figure 14-13. Expressing −5 in 2's Complement

Another example of 2's complement is shown in Figure 14-14 with the steps required to express −7 in 2's complement.

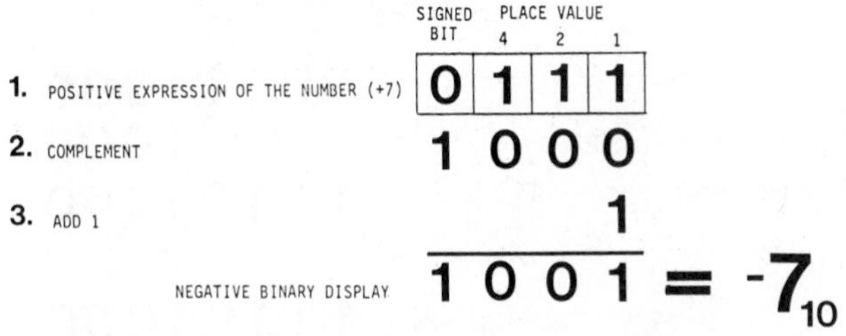

SIGNED BIT   PLACE VALUE   4   2   1

1. POSITIVE EXPRESSION OF THE NUMBER (+7) $\boxed{0\;|\;1\;|\;1\;|\;1}$

2. COMPLEMENT   1 0 0 0

3. ADD 1   1

NEGATIVE BINARY DISPLAY $1\;0\;0\;1 = -\mathbf{7}_{10}$

Figure 14-14. Expressing −7 in 2's Complement

To convert a negative binary display to the decimal equivalent, the negative binary display is complemented; 1 is added; the binary sum is converted to decimal and the negative sign is added.

Figure 14-15 shows the steps to determine what negative value 1110 represents.

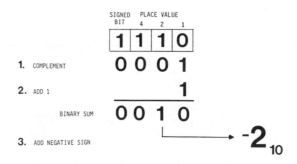

Figure 14-15. 2's Complement to Decimal Equivalent

Another method of converting positive numbers to 2's complemented negative numbers is as follows.

Starting at the least significant bit and working to the left, copy each bit up to and including the first 1 bit and then complement or change each remaining bit. How to express −2 in 2's complement using a 4 bit word and this alternate method is shown in Figure 14-16.

1. ORIGINAL POSITIVE NUMBER (+2)   0010
2. COPY UP TO FIRST BIT   10
3. COMPLEMENT THE REMAINING BITS   1110

2'S COMPLEMENT -2   1110

Figure 14-16. Alternate Method of 2's Complement

Another example is shown in Figure 14-17 for 2's complementing −24 using an 8 bit word.

1. ORIGINAL POSITIVE NUMBER (+24)   00011000
2. COPY UP TO FIRST 1 BIT   1000
3. COMPLEMENT REMAINING BITS   11101000

2'S COMPLEMENT -24   11101000

Figure 14-17. Complement of −24 Decimal

By using 2's complement, negative and positive values can now be added.

The two steps for adding -7_{10} and $+5_{10}$ using 2's complement with a 4 bit word is shown in Figure 14-18.

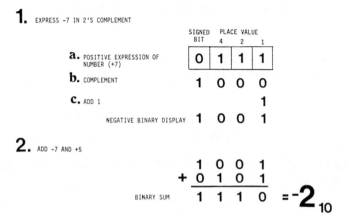

Figure 14-18. Adding Positive and Negative Numbers

NOTE: When adding signed binary numbers, any carry from the signed bit column is discarded.

Square D SY/MAX Model 300 uses a 16 bit word format for arithmetic functions, with bit 16 being designated as the signed bit. The words used for the arithmetic functions may be values stored in registers, constants, or combinations of registers and constants.

If a LET statement is programmed to perform the addition in the previous example, the rung could appear as shown in Figure 14-19.

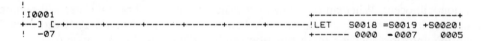

Figure 14-19. Programmed Addition of Negative and Positive Numbers

NOTE: To enter a negative value into a storage register with the SY/MAX 300, place a zero (0) after the equal sign and subtract the required number from it, that is, let S19 = 0 −7. This programming technique will enter −7 into register S19.

When input device 01-07 is closed or goes TRUE, the values stored in registers S19 and S20 (−7 and +5) will be added and the sum (−2) stored in register S18.

Using the data mode of the SY/MAX CRT, the register display for registers 18, 19, and 20 (S18, S19, and S20) would be as indicated in Figure 14-20.

	DECIMAL	HEX	16... .BIN/FLP. ...1	32... .STATUS.. ...17
S0019	-0007	FFF9	1111 1111 1·111 1001	0001 0011 0000 0000
S0020	0005	0005	0000 0000 0000 0101	0001 0011 0000 0000
S0018	-0002	FFFE	1111 1111 1111 1110	0001 0011 0000 0000

Figure 14-20. CRT Display of Registers Showing
Decimal, Hexidecimal, and Binary Numbers

Once addition of signed numbers is possible, the other arithmetic functions — subtraction, multiplication, and division — are also possible, because with a PC, they are achieved by successive addition.

For example, subtraction of the number 20 from 26 is accomplished by complementing 20 to obtain −20 and then performing addition.

Subtracting 20 from 26 by complementing 20 and performing addition using an 8 bit word is shown in Figure 14-21.

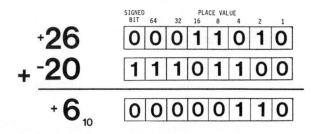

Figure 14-21. Subtraction by Addition

The Square D SY/MAX 300 PC limits any arithmetic operation answer to plus or minus 32,767 — the amount that can be displayed in one 16 bit word where bit 16 is used as a sign bit. If the results of any arithmetic operation exceeds plus or minus 32,767, an overflow condition will occur. When an overflow exists, the data in the storage register will remain the same as it was prior to the operation. When only one register or storage word is used for storing and/or displaying math function results, it is referred to as single (one) precision arithmetic.

The four function arithmetic feature (+, −, ×, and ÷) can be used with constant values or values stored in a storage register, holding register, input/output registers, or any other accessible word locations.

A typical application of an arithmetic function could be a chemical batch plant where a given mix of two chemicals (A and B) is to have a 2:1 ratio. By using analog input devices as discussed in Chapter 2, the weight of chemical A can be converted to a binary equivalent and stored in word 40. For a 2:1 mix, only one half as much of chemical B by weight is used.

The binary value in storage word 40 can be divided by two (2) to determine the amount of chemical B that should be used for a proper 2:1 mix.

Figure 14-22, using a Square D LET instruction, shows how the value in storage word S40 can be divided by 2 and the result placed in storage word S41. A data comparison could then be made to the data in storage word S41 to limit the amount of chemical B to one half the amount of chemical A.

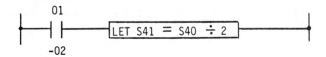

01

-02

Figure 14-22. Square D Arithmetic Instruction

When 01-02 is closed, the value in storage word 40 (S40) will be divided by 2 and the result stored in storage word 41 (S41) on each processor scan.

The arithmetic functions might also be used with timer/counter values, accumulated or preset, to change a given process machine operation under varying conditions.

Using a Square D LET instruction, the basic arithmetic functions used with storage words and constants are shown in Figures 14-23a, b, c, and d.

Ø1

-Ø8

Figure 14-23a. Square D Add Instruction

When 01-08 is true, a constant of 003 will be added to the value in storage word 20 and the sum stored in storage word 15.

Ø1

-Ø7

Figure 14-23b. Square D Subtract Instruction

When 01-07 is true, the value in storage word 4 will be subtracted from a constant of 1000 and the difference stored in storage word 40.

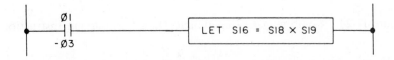

Ø1

-Ø3

Figure 14-23c. Square D Multiply Instruction

205

When 01-03 is true, the value in storage word 18 is multiplied by the value of storage word 19 and the product stored in storage word 16.

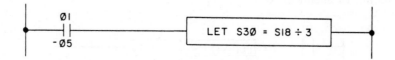

Figure 14-23d. Square D Divide Instruction

When 01-05 is true, the value in storage word 18 will be divided by 3 and the result stored in storage word 30.

Arithmetic functions can also be used with Square D IF data comparison instructions as shown in Figure 14-24.

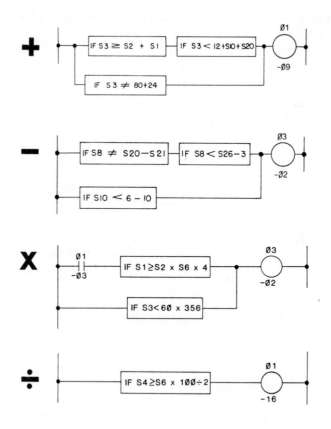

Figure 14-24. Combining Arithmetic and Data Comparison Instructions

206

The addition and subtraction instructions for the Allen-Bradley PC are used in conjunction with the GET instruction and are other examples of single precision arithmetic where only one word is used to store the results of either addition or subtraction.

Figure 14-25 shows two GET instructions followed by the ADD instruction. When input device 11105 is true, the values stored in words 030 and 031 are added and the sum stored in word 032. Allen-Bradley arithmetic functions limit the number in any storage word to 999 BCD. If the results of the addition exceed 999, bit 14 of the addition storage word (032) will be set to 1 and a 1 will be displayed on the industrial terminal.

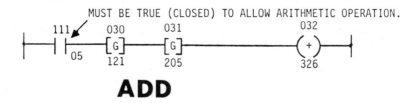

ADD

Figure 14-25. Allen-Bradley Add Instruction

The subtraction instruction also uses two GET instructions (050 and 062) with a SUBTRACTION instruction (071) for single precision arithmetic as shown in Figure 14-26. If the result of the subtraction is a negative number, bit 16 of the subtraction storage word (071) will be set to 1 and a negative sign (−) will be displayed on the industrial terminal.

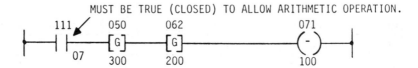

SUBTRACT

Figure 14-26. Allen-Bradley Subtract Instruction

Figure 14-27 shows yet another variation of single precision arithmetic using the Westinghouse PC-700/900 format.

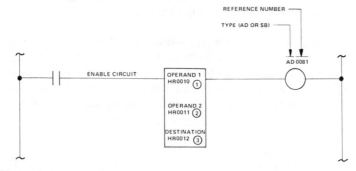

Figure 14-27. Westinghouse Add and Subtract (AD/SB) Instruction

207

When the enable circuit is true, operand 1 (HR0010) and operand 2 (HR0011) are added and the result is placed in the destination register (HR0012). If the result is equal to or greater than 10,000, 10,000 is subtracted and coil AD0081 (add coil) is energized. The coil being energized acts as an overflow alert to indicate that a value equal to or greater than 10,000 has been reached. Any value over 10,000 is then stored in the destination register HR0012.

For subtraction, operand 2 is subtracted from operand 1 and the result placed in the designation register. If the result is less than 0000, coil SB (subtract) is energized and the amount of underflow (amount less than 0000) is placed in the destination register.

By using a technique called double precision arithmetic, two registers or storage words are used to store the results of arithmetic operations, adding to the maximum value that can be stored.

Figure 14-28 shows a block diagram of double precision addition using the General Electric series six format. When input device I0001 is true, the value stored in references or registers R0002 and 0001 will be double precision added (DPADD) to references R0004 and R0003 with the sum stored in registers R0006 and R0005. With this format, numeric values from a maximum of $+2,147,483,647$ to a minimum of $-2,147,483,648$ may be used. If the results of the addition is greater than $+2,147,483,647$ registers R0006 and 0005 will contain the value and output O 0001 will be turned ON. If the sum of the addition is less than $-2,147,483,648$ again registers R0006 and 0005 will hold the value and output O 0001 will be ON. If the results are between the two limits, the value will again be stored in registers R0006 and 0005, but output 00001 will be turned OFF.

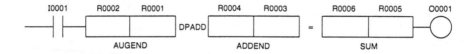

Figure 14-28. Block Diagram G.E. Series Six Double Precision Addition

While the limits and limitations of the different arithmetic functions vary from manufacturer to manufacturer, the concepts are basically the same. Values or contents of storage registers, data table words, or constants are combined arithmetically and the results stored in register or data table words. While the actual arithmetic function is performed using binary 2's complement, the number values may be stored and/or displayed in binary/BCD, hexadecimal, or octal depending on the PC circuitry.

The divide and multiply instructions for Allen-Bradley PC's use two data tables, or storage words, to store the double precision results of both division and multiplication. The product of multiplication is limited to 998,001 (999 × 999) while the quotient of the division instruction is limited to 3 whole numbers and 3 decimal places. Figures 14-29 and 14-30 are examples of programmed multiply and divide instructions using the Allen-Bradley format.

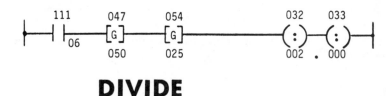

MULTIPLY

Figure 14-29. Allen-Bradley Multiply Instruction

When input device 11110 is true (closed), the value of word 033 (150) is multiplied by the value in word 041 (010), and the product is stored in words 050 and 051 (001 and 500 or 1500).

NOTE: Two memory words are used so that a number larger than 999 can be displayed and stored.

DIVIDE

Figure 14-30. Allen-Bradley Multiply Instruction

When input device 11106 is true (closed), the value in word 047 (050) will be divided by the value in word 054 (025), and the result is stored in words 032 and 033 (002.000 or 2).

As previously discussed, the only way to learn how to apply the arithmetic functions on a given PC is to read the manufacturer's literature and work with that PC.

Chapter Summary

For programmable controllers to perform arithmetic functions, a way must be found to represent both positive and negative numbers. One of the most common methods used is called 2's complement. With 2's complement, negative and positive numbers can be added, subtracted, divided, and multiplied. In reality, however, all arithmetic functions are accomplished by successive addition. As with most other features, arithmetic functions and format will vary with each manufacturer. Arithmetic functions of add, subtract, multiply, and divide can be combined with data manipulation instructions (data transfer and data compare) to provide expanded control for and information from process or driven equipment. Memory words such as holding, storage, and data can be used with the arithmetic functions as well as words and constants or just constants.

Review Questions

1. List the four math functions that can be performed by many programmable controllers.

   Show the steps used to express the following signed decimal numbers in 2's complement. Use 8 bit words.

2. -7

3. -4

4. -3

   Convert the following decimal numbers to 2's complement and add. Show all work. Use 8 bit words.

5. $\begin{array}{r} +4 \\ \underline{-7} \end{array}$

6. $\begin{array}{r} -10 \\ \underline{+22} \end{array}$

7. $\begin{array}{r} +22 \\ \underline{+33} \end{array}$

8. T   F   Data manipulation instructions can be combined with arithmetic instructions.
9. What is double precision math used for?
10. Give an example of how the arithmetic instructions could be used.

Chapter 15

Word and File Moves

Objectives

After completing this chapter, you should have the knowledge to
- Describe the function of a synchronous shift register.
- Understand the function of word to file, file to word, and file to file instructions.
- Explain the difference between an asynchronous shift register (FIFO) and word to file move.

Before word and file moves are discussed, understanding the definition of both words and files will be helpful.

Words, or registers as they are often referred to, are locations in memory that can be used to store different kinds of information. Typically a word or register can store the status of inputs and outputs, hold numerical values used for math functions or other numerical data used for timers, counters, etc.

A **file** is a group of consecutive memory words used to store information. Words 1 through 5 would make up a consecutive 5 word file. Words 1, 2, 3, 6, and 7 could not be used as a 5 word file because the numbers are not consecutive.

A file may also be referred to as a table.

WORDS

Information stored in a word can be shifted within the word, or from one word to another word. Information stored in a word may also be moved into a file or the information stored in a file can be transferred into a word. All of these different possibilities will be discussed later in this chapter.

SYNCHRONOUS SHIFT REGISTER

When information is shifted one bit at a time within a word or from one word to another, it is called a synchronous shift register. The bits may be shifted forward (left) or reverse (right).

NOTE: The synchronous shift register may also be referred to as a serial shift register.

Figure 15-1 shows a 16 bit word used as a forward synchronous shift register.

a.

16	15	14	13	12	11	10	9	8	7	6	5	4	3	2	1
0	0	1	1	1	0	0	1	1	1	0	0	1	1	1	0	Register 100

Initial Condition of Register

b.

| 0 | 1 | 1 | 1 | 0 | 0 | 1 | 1 | 1 | 0 | 0 | 1 | 1 | 1 | 0 | 0 | Register 100

Condition Register After Shift Forward (Left)

Figure 15-1. Forward 16 Bit Synchronous Shift Register

Figure 15-1a shows the bit status of a register (word) 100 prior to the forward shift, while 15-1b shows how the register would look after the bits have been shifted one place to the left or forward.

Notice that when the register was shifted, the information (1 or 0) in bit 16 is shifted out or is lost. If the register was continually shifted with a zero (0) in bit location 1, all of the 1's would be shifted left or forward until only zeros remain (Figure 15-2).

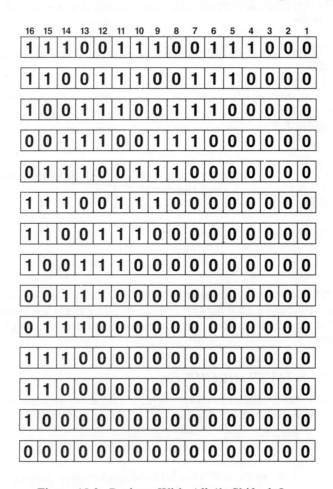

Figure 15-2. Register With All 1's Shifted Out

In a forward shift register, bit 1 is used to enter data (1 or 0) and then the data is shifted forward one bit at a time. Figure 15-3 shows a two word forward shift register. In this case, data is entered at bit 1 and shifted one bit at a time to the left. With a two word shift register, the information (1 or 0) in bit 16 of word 1 is not shifted out and lost, but is shifted into bit 1 of the second word of the shift register.

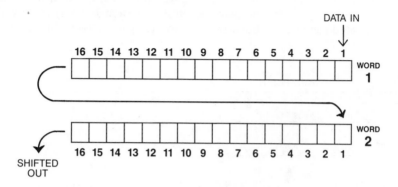

Figure 15-3. Two Word Forward Shift Register

Figure 15-4 shows the Square D format for a shift register.

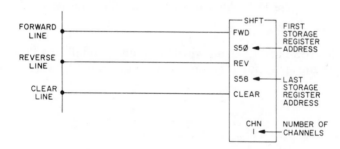

Figure 15-4. Square D Format for a Synchronous Shift Register
(Courtesy of Square D Company)

The shift register consists of a forward line, reverse line, a clear line, and a shift box that contains three pieces of information: the first storage register address to be used in the shift register, the last storage register address (would be the same as the first storage register if the shift register only used one 16 bit word), and the number of channels. The channel specifies the number of bits to be shifted at one time. The choices are 1 as discussed, 8, or 16.

An eight channel shift register would shift 8 bits at a time either forward or reverse. A 16 channel shift register would shift all 16 bits from one word to another.

The forward line will shift the data in the register(s) forward (left) on each false to true (open to close) transition.

The reverse line will shift the data in the register(s) reverse (right) on each false to true transition.

The clear line enables the shift box when there is continuity. When the clear line is open, the data in the register(s) will be set to 0 and transitions of either the forward or reverse lines will have no effect.

What might be a practical application for a shift register? Consider the overhead parts conveyer in Figure 15-5 that is used to transport parts into a paint booth for painting. If a part is on the hook as it enters the paint booth limit switch 1 (LS-1) is activated. Limit switch 2 is activated each time a hook on the conveyer passes, even if no part is present.

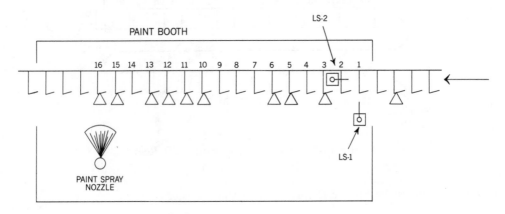

Figure 15-5. Applying a Forward Shift Register

Both limit switches (LS-1 and LS-2) are wired to an input module of a PC and the solenoid that operates the paint spray nozzle is wired to an output module as shown in Figure 15-6.

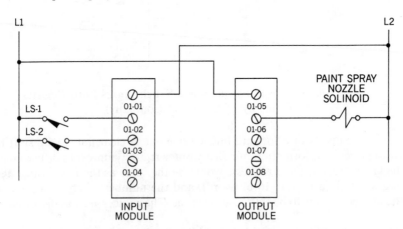

Figure 15-6. Input and Output Devices Wired to I/O Modules

The input addresses of the limit switches and the output address of the paint spray nozzle can now be programmed with a forward shift register as shown in Figure 15-7.

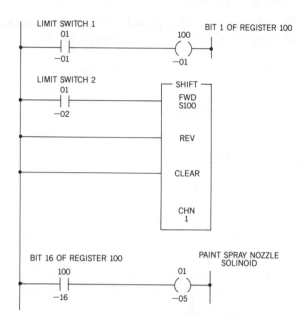

Figure 15-7. Programming a Forward Shift Register Using Square D Format

Notice in the shift box of Figure 15-7, that storage word S100 is being used as the shift register.

When a part activates LS-1 (address 01-01) and the limit switch closes, rung 1 goes true and a 1 is placed in bit 01 of word 100. As the part moves toward the spray nozzle, LS-2 is activated by the hook which closes LS-2. Closing of LS-2 contacts (address 01-02) gives a false to true transition in the foward (FWD) line and the shift register shifts the information in word 100 one place to the left. The shift moved the 1 in bit 1 to the bit 2 location. As the conveyer continues to run, a 1 or 0 is entered into bit 1 or word 100 depending on whether a part is present or not. The data is then shifted as LS-2 is activated and deactivated by the moving hooks. As the data shifts to the left, the paint spray nozzle solenoid in rung 3 is activated each time a 1 is shifted into bit 16 of word 100.

A reverse shift register is shown in Figure 15-8. Note that with a reverse shift the data enters at bit 16 and is shifted out at bit 1.

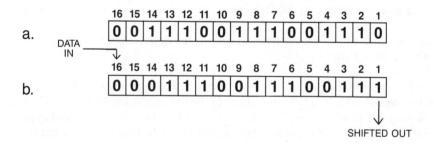

Figure 15-8. Reverse (Right) Shift Register

Like other functions, shift register formats will vary from manufacturer to manufacturer, but the basic function of the synchronous shift register will be the same.

FILE MOVES

As indicated earlier, a file is a group of consecutive words used to store or hold information. A file can consist of just a few words or can be several hundred words in length depending on the PC.

Figure 15-9 shows a five word file using consecutive memory words 50 through 54.

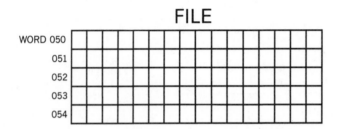

Figure 15-9. Five Word File

Information (data) may be transferred into or out of a file by using data transfer instructions. The three most common data transfer instructions are:
1. Word to file
2. File to word
3. File to file

Since each PC manufacturer will use a different instruction set and keystrokes to implement these instructions, the different concepts will only be discussed in general.

WORD TO FILE INSTRUCTION

The word to file instruction is used to transfer data from a word into a file. For illustration, let's assume that word 110 stores the temperature of the die for a plastic injection molding machine. A thermocouple is attached to the heated die and then connected to a thermocouple input module. Depending on the module, the temperature of the die is then stored in an input word in either binary or BCD format.

By using a word to file data transfer instruction, the data (temperature) in word 110 can be transferred into a file.

Once the word to file instruction has been programmed, the information stored in word 110 will be transferred into a file each time the instruction is implemented. Figure 15-10a shows a 5 word file prior to a word to file instruction being implemented, and Figure 15-10b shows the file after the data transfer instruction (word to file) has been implemented.

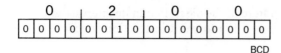

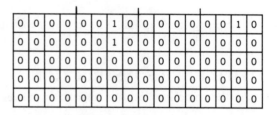

Figure 15-10a. File Prior to Word to File Data Instruction Implementation

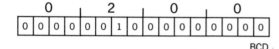

Figure 15-10b. File Content After First Word to File Instruction is Implemented

The next time the word to file instruction is implemented (indexed), the current value in word 110 will be transferred to the file. All information currently in the file will be moved down to make room for the new data. Figure 15-11 illustrates this point.

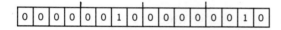

Figure 15-11. File After Second Word to File Instruction is Implemented

By using a timer, the word to file instruction could be implemented every 15 minutes. By increasing the size of the file, a record of the die temperature for an eight hour shift could be stored. The data (temperature) from the file could be printed out and the temperature of the die compared to quality control records. The application of this instruction, like all other instructions, is only limited by imagination.

FILE TO WORD INSTRUCTION

As you might suspect, the file to word instruction transfers data from a file into a word.

Using the previous example, the temperature of the injection molding machine die could be transferred to an output word (011) that controls a LED display. By incrementing or indexing the file to word instruction, the temperature of the die in 15 minute intervals could be displayed. Figures 15-12a and b illustrate how a file to word instruction functions.

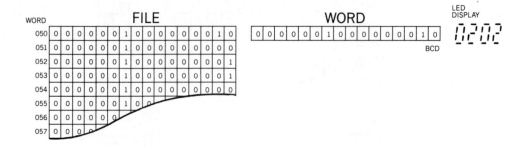

Figure 15-12a. File to Word Instruction at First Word of File (050)

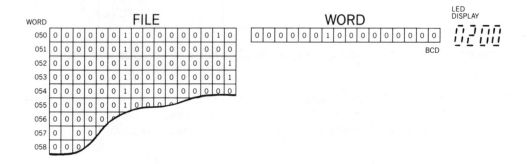

Figure 15-12b. File to Word Instruction at Second Word of File (051)

FILE TO FILE INSTRUCTION

As the name implies, this instruction moves data from one file to another file.

An excellent example for using the file to file move would be a chemical batch plant where different amounts and types of chemicals are mixed for a variety of products.

The different mix ratios (recipes) would be stored in different files and would be transferred to a file that controls machine and/or plant operation for a given product.

The file to file move is similar to a GET-PUT or LET instruction, except rather than just transferring data from one word to another, the file to file move transfers data from several words at one time. Figure 15-13 shows how a file to file move would work.

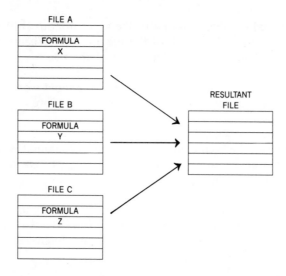

Figure 15-13. Data in File A, B, or C Can Be Transferred Into
the Resultant File When the File to File Move is Executed

ASYNCHRONOUS SHIFT REGISTER (FIFO)

The asynchronous shift register rather than shifting bits of information within a word or words like the synchronous shift register, shifts the data from a complete word into a file or stack. While this appears to be just another name for a word to file instruction, it is not. While there are similarities between the two, there is one major difference.

Remember that in a word to file move that the information from the word was shifted into the top of the file and moved down through the file with each implementation or indexing of the instruction. The asynchronous shift register allows the information transferred from a word to go to the last unused word of the file. This difference is why the asynchronous shift register is often referred to as a FIFO stack—first in/first out.

Figure 15-14 compares both the asynchronous shift register and the word to file instructions to demonstrate the difference.

ASYNCHRONOUS SHIFT REGISTER		WORD TO FILE MOVE	
INPUT WORD FILE or STACK		INPUT WORD FILE	

a. 0 0 1 1 | 0 0 0 0 / 0 0 0 0 / 0 0 0 0 / 0 0 1 1 | 0 0 1 1 | 0 0 1 1 / 0 0 0 0 / 0 0 0 0 / 0 0 0 0

b. 0 0 0 0 | 0 0 0 0 / 0 0 0 0 / 0 0 0 0 / 0 0 1 1 | 0 0 0 0 | 0 0 0 0 / 0 0 1 1 / 0 0 0 0 / 0 0 0 0

c. 1 0 1 1 | 0 0 0 0 / 0 0 0 0 / 1 0 1 1 / 0 0 1 1 | 1 0 1 1 | 1 0 1 1 / 0 0 0 0 / 0 0 1 1 / 0 0 0 0

Figure 15-14. Comparison of Asynchronous Shift Register (FIFO) to Word to File Move

219

When data is transferred at (a) notice that in the asynchronous shift register the data is placed in the last (bottom) unused word of the file or stack. Whereas in the word to file move, data is transferred into the first (top) word of the file.

At the next implementation (b) no data is present in the input word, so no change takes place in the FIFO stack but note that the previous data is moved down one word in the word to file instruction.

When the instructions are indexed again (c), the data of the input word is transferred to the next available word at the bottom of the FIFO stack, whereas the data is entered at the top of the file and all previous data is shifted down in the word to file instruction.

The big difference then between the two instructions is the ability of the word to file move to accept and store all zeros while the asynchronous shift register (FIFO) ignores input data of all zeros. The FIFO stack is useful when you are only interested in data and are not concerned with periods when no data is transferred.

Chapter Summary

While the keystrokes and instruction will vary with each PC manufacturer, the principles of word and file moves are the same. The synchronous shift register shifts bits of information left or right (forward and reverse) within a word or words, while file moves can transfer data from words to files, files to words or files to files. The asynchronous shift register is referred to as a FIFO, or first in/first out, as data transfers or falls to the bottom of the stack and uses the last unused word.

Review Questions

1. Define the term *word* as used in this chapter.
2. T   F   The synchronous shift register shifts data in a forward direction only.
3. In a one word shift register, the data is entered as bit:
   a. 1
   b. 2
   c. 4
   d. 8
   e. 16
   f. None of the above
4. In a one word shift register, the data is shifted out at bit:
   a. 1
   b. 2
   c. 4
   d. 8
   e. 16
   f. None of the above

5. Define the term *file* as used in this chapter.
6. Which of the following groups of words could NOT be a file:
    a. 50, 51, 52
    b. 50, 51, 52, 53
    c. 100, 101, 102, 103
    d. 100, 101, 102, 103, 105
7. Briefly describe the function of a word to file move.
8. Briefly describe the function of a file to word move.
9. Briefly describe the function of a file to file move.
10. Which instruction is also known as a FIFO (first in/first out):
    a. Synchronous shift register
    b. Word to file move
    c. File to word move
    d. Asynchronous shift register
    e. File to file move
11. T   F   When data is transferred into a file using a word to file move, the data is entered at the last unused word of the file.

Chapter 16

Sequencers

Objectives

After completing this chapter, you should have the knowledge to
- Describe what a sequencer instruction does.
- Understand the basics of sequencer operation.
- Define the term *mask*.

Sequencers are used to transfer information from memory words to output addresses for the control of sequential machine operation.

A programmed sequencer will replace the mechanical drum sequencer that has been used in the past. On the mechanical sequencer, when the drum cylinder was rotated, contacts were caused to open and close mechanically to control output devices. Figure 16-1 shows a mechanical drum cylinder with pegs placed at varying horizontal positions for Step 1 of the sequence. When the cylinder is rotated, contacts that align with the pegs will close, while the contacts where there are no pegs will remain open. In this example the presence of a peg could be thought of as a 1 or ON, while the absence of a peg would be 0 or OFF.

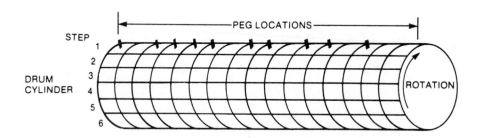

Figure 16-1. Drum Cylinder
(Courtesy of Allen-Bradley)

To program a sequencer, binary information is entered into a series of consecutive memory words. These consecutive memory words are referred to as a **file**. Information, from the words in the file, is then transferred sequentially to the output word to control the outputs.

If the first six steps on the drum cylinder in Figure 16-1 were removed and flattened out, they would appear as illustrated in Figure 16-2.

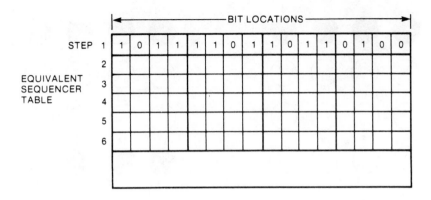

Figure 16-2. Sequencer Table

For Step 1, each horizontal location where there was a peg is now represented by a 1 (ON), and the positions where there were no pegs are represented by a 0 (OFF).

The six steps could also be viewed as a 6 word file with each 16 bit word representing a sequencer step. By entering different binary information, 1's and 0's, into each word of the file, the file can replace the rotating drum cylinder.

To illustrate how this works, 16 lamps for outputs as shown in Figure 16-3 will be used.

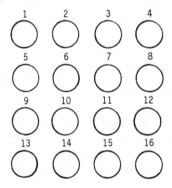

Figure 16-3. Output Lamps

Each lamp represents one bit address (1 through 16) of output word 25.

Assume, for the sake of discussion, that the operator wants to light the lamps in the four step sequence shown in Figures 16-4a, b, c, and d.

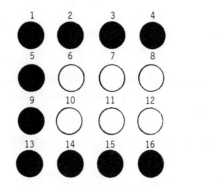

Figure 16-4a. Sequencer Step 1

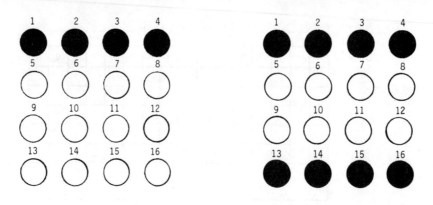

Figure 16-4b. Sequencer Step 2

Figure 16-4c. Sequencer Step 3

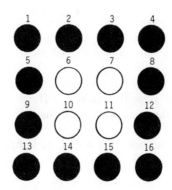

Figure 16-4-d. Sequencer Step 4

The bit addresses of the lamps that are to be lit in each step of the sequence should be written down to assist in entering data into a word file.

The next step is to define a word file to store the binary data required for each step of the sequencer. Words 30, 31, 32, and 33 can be used for the four word file. By using the programmer, binary information, 1's and 0's, can be entered into each word of the file to reflect the desired lamp sequence as indicated in Figure 16-5.

	16	15	14	13	12	11	10	9	8	7	6	5	4	3	2	1	
WORD 25	0	0	0	0	0	0	0	0	0	0	0	0	0	0	0	0	OUTPUT
WORD 30	0	0	0	0	0	0	0	0	0	0	0	0	1	1	1	1	STEP #1
WORD 31	1	1	1	1	0	0	0	0	0	0	0	0	1	1	1	1	#2
WORD 32	1	1	1	1	0	0	0	1	0	0	0	1	1	1	1	1	#3
WORD 33	1	1	1	1	1	0	0	1	1	0	0	1	1	1	1	1	#4

Figure 16-5. Binary Information for Each Sequencer Step

224

NOTE: Some PC's allow the data to be entered using BCD which speeds the entry process. To use this feature, the required binary information for each sequencer step is converted to BCD as covered in Chapter 4. The information is then entered with the programmer into the word file with four key strokes for each word, rather than 16.

Once the sequencer has been programmed and the data entered into the word file, the seqencer is ready to control the lamps. When the sequencer is activated and advanced to Step 1, the binary information in word 30 (Figure 16-5) is transferred into word 25, and the lamps will light in the pattern shown in Figure 16-4a. Advancing the sequencer to Step 2 will transfer the data from word 31 into word 25 for the light sequence shown in Figure 16-4b. Step 3 transfers the data from file word 32 into word 25, while Step 4 transfers information from word 33. When the last step is reached, the sequencer can be reset and sequenced again.

Since each PC with sequencer capabilities is programmed differently, no attempt will be made to explain the actual programming. Depending on the PC, sequencers can be programmed from a few steps up to hundreds of steps and can control only one output word or several.

MASKS

When a sequencer operates on an entire output word, there may be outputs associated with the word that the operator does not want controlled by the sequencer. To prevent the sequencer from controlling certain bits of an output word, a **mask** word is used. Figure 16-6 shows pictorially how a mask word works.

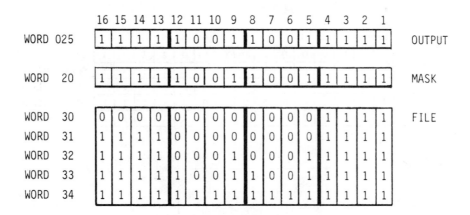

Figure 16-6. Using a Mask Word

The mask word is a means of selectively screening out data from the sequencer word file to the output word. For each bit of output word 025 that the operator wants the sequencer to control, the corresponding bit of mask word 20 must be set to 1.

In Figures 16-4a, b, c, and d bits 6, 7, 10, and 11 are not used. By not setting bits 6, 7, 10, and 11 of the mask word to 1, these bits can be used independently of the sequencer.

In Figure 16-6 a fifth step was added to the sequencer, file word 34, and bits 6, 7, 10, and 11 are set to 1. With mask word 20 having bits 6, 7, 10, and 11 set to 0, the data in file word 34 is screened out and prevented from being transferred into output word 025.

You have probably noticed that the sequencer works much like the file to word move discussed in Chapter 15. For programmable controllers that don't have a dedicated sequencer instruction, a file to word move instruction can be used.

Another method of creating a sequencer when no sequencer or word to file move is available, is to use a 16 channel synchronous shift register.

Remember from the previous chapter that a typical synchronous shift register can consist of one or more words. Also, the shift register can shift one bit at a time, 8 bits or 16 bits (one word) at a time.

By using a 16 channel (16 bit shift) multiword shift register we can create a sequencer.

A 16 channel shift register with 5 words can be used to duplicate the sequencer steps shown in Figure 16-5. Remember that the synchronous shift register shifts new information in on each shift and likewise shifts information out on each shift.

To recycle the information and get the cylindrical (repeating) action of a sequencer will require the use of data transfer instructions.

Figure 16-7 shows a five word file and a Square D SY/MAX 16 channel synchronous shift register instruction that will be used for the sequencer. Notice that two data transfer instructions are also used.

BITS

	16	15	14	13	12	11	10	9	8	7	6	5	4	3	2	1	WORD	
DATA IN	0	0	0	0	0	0	0	0	0	0	0	0	1	1	1	1	0017	
	1	1	1	1	1	0	0	1	1	0	0	1	1	1	1	1	0018	STEP 4
	1	1	1	1	0	0	0	1	0	0	0	1	1	1	1	1	0019	STEP 3
	1	1	1	1	0	0	0	0	0	0	0	0	1	1	1	1	0020	STEP 2
DATA OUT	0	0	0	0	0	0	0	0	0	0	0	0	1	1	1	1	0021	STEP 1

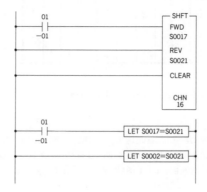

Figure 16-7. Programming a 16 Bit Shift Register as a Sequencer

When input 01-01 is closed, the shift register will shift forward at the same time the first data transfer rung will be true. The data transfer says: LET S0017 = S0021 or take the data in register 0021 and put it in register 0017. This rung has the effect of recycling the data. Normally, when the shift register is shifted forward, the data in register 0021 would be shifted out and lost. Now on each shift the information from register 0021 is transferred up and into register 0017. On the next shift, the data in 0017 is shifted down to register 0018 and the data from register 0021 is again shifted into register 17.

The second data transfer rung is programmed unconditional (true all the time) and will transfer the data from register (word) 0021 into output word (register) 0002 to obtain the required light display.

The binary information for each sequencer step is entered into each register by using the data mode on the Square D SY/MAX programmer. While the Square D format was used for this example, what the circuit really illustrates is that by knowing and understanding PC functions unique programming is possible.

Chapter Summary

Sequencers, like other data manipulation and arithmetic instructions, are programmed differently with each PC, but again the concepts are the same. Data is entered into a word file for each sequencer step, and, as the sequencer advances through the steps, binary information is transferred sequentially from the word file to the output word(s). Output word bits can be masked so they can operate independently of the sequencer.

Review Questions

1. Briefly describe a sequencer.
2. A series of consecutive words are referred to as
   a. Deck
   b. Group
   c. File
   d. Chain
3. What is the purpose of a mask word in a sequencer?
4. Set up the file in Figure 16-A so the sequencer will operate the motors as shown in Steps 1, 2, 3, and 4, and motors 01014, 01015, 01016, and 01017 so they cannot be energized.

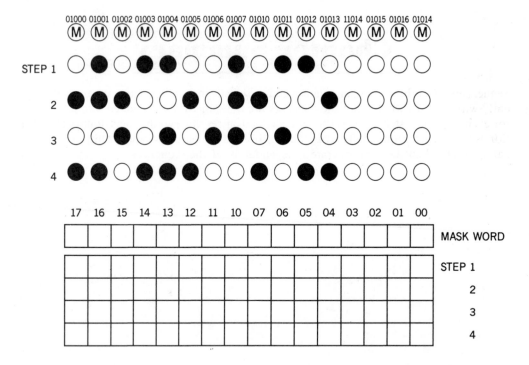

Figure 16-A

Chapter 17

Start-Up and Troubleshooting

Objectives

After completing this chapter, you should have the knowledge to
- Understand start-up procedures listed in manufacturers' literature.
- Explain how input devices are tested.
- Explain how to test output devices using a pushbutton or other input device.
- Explain safety consideration when testing output devices.
- Describe how voltage readings are taken to check input and output modules.

START-UP

Careful start-up procedures are necessary to prevent damage to the driven equipment, the programmable controller system, or more importantly, injury to personnel.

Prior to beginning a system start-up procedure, it is important to check and verify that the system has been installed according to the manufacturer's specifications and also that the installation meets local, state, and national codes. Pay special attention to system grounding.

Before applying power to the controller, the following steps should be completed.
1. Verify that the incoming power matches the jumper selected voltage setting of the power supply. Figure 17-1 shows a typical AC power terminal strip and has the jumper position indicated at the sides of the strip.

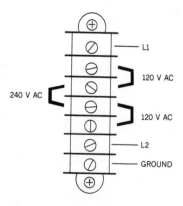

Figure 17-1. Power Supply Terminal Strip with Jumper Positions Indicated

NOTE: For 120V AC the neutral conductor is connected to the L2 terminal and the 240V jumper is removed.

2. Verify that a hard wire safety circuit or other redundant emergency stop device as described in Chapters 2 and 10 has been installed and is in the open position.
3. Check all power and communication cables to ensure connector pins are straight, not bent or pulled out.
4. Connect all cables making sure that connectors are fully inserted into their sockets. Secure connectors as applicable.
5. Ensure all modules are securely held in the I/O rack and that field wiring arms (if applicable) are fully seated and locked.
6. Place PC processor key switch to a safe position as indicated in Figure 17-2.

Figure 17-2. Key Switch in Halt (Safe) Position

7. Double check the setting(s) of all DIP switches.

CAUTION: Before proceeding further make sure that the safety circuit or other emergency stop device is OFF or OPEN and that power is removed from all output devices.

Apply power and observe processor indicator light(s) for proper indications.

With power applied and the power supply providing the necessary DC voltage for the processor and I/O rack, the input indicator LED's of the input modules will be functioning. Any input device that is closed, or ON, will have an illuminated LED (Figure 17-3).

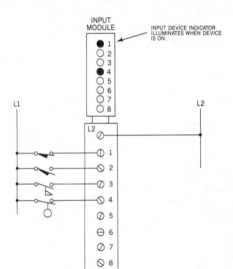

Figure 17-3. Input Module Indicators

230

TESTING INPUTS

Each input device can be manipulated to obtain open and closed contact conditions.

CAUTION: Do NOT activate the input devices mounted on equipment by hand; unexpected machine motion could cause injury. Use a wooden stick or other nonconducting material to activate input devices mounted on equipment.

Each time an input device is closed, the corresponding LED on the input module should illuminate. Failure of a LED to illuminate indicates:

1. Improper input device operation.
2. Incomplete or incorrect wiring (check that the input device is wired to the correct input module and proper terminal).
3. Loss of power to the input device.
4. Defective LED and/or input module.

As a further check of the system, a program can be developed and entered into the processor that uses each input device address. With the key switch in the test or disable output position, the status of the input devices may now be monitored on the CRT of a desktop programmer or by LED indicators on a hand held programmer. On a CRT, the input contact will become intensified or go to reverse video when the instruction is true. An EXAMINE ON instruction would intensify or go to reverse video when the input device was ON or closed. An EXAMINE OFF instruction would be intensified or show reverse video when the input is open or OFF. Figure 17-4 shows a rung for testing input devices.

Figure 17-4. Rung for Testing Input Devices

Once all input devices have been tested and checked out as being operational and properly terminated, the output devices can be tested.

TESTING OUTPUTS

Before testing output devices, it must be determined which devices can safely be activated and which devices should be disconnected from the source of power. Figure 17-5 shows a motor starter with the motor disconnected for safety. In this configuration, the motor starter coil (the output device) can be activated for check-out without energizing the motor. This prevents unwanted machine motion that could cause injury to personnel or machine damage.

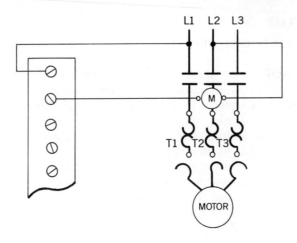

Figure 17-5. Disconnecting Motor Leads for Safety

For outputs that can safely be started, be sure equipment is in the start-up position and is properly lubricated and ready to run.

There are two methods that may be used to test output devices. The first method uses a pushbutton or other convenient input device that is part of the control panel. The pushbutton will be programmed to energize each output, one at a time.

The second method will use the FORCE function of the PC to energize outputs one at a time.

When using a pushbutton (or other input device), the address of the pushbutton is programmed in series with the output device you wish to test (Figure 17-6).

Figure 17-6. Rung for Testing Output Device

Once the rung has been programmed and entered into processor memory, the processor is placed in the run mode. Pressing the pushbutton should illuminate the output indicator on the output module for address 01000. If there is an output connected, verify that the output device is energized.

If the output indicator does not illuminate, verify that the input instruction 11105 is intensified or showing reverse video to indicate an ON condition. Double check the output address and verify that the output instruction indicates an ON condition. If both instructions indicate an ON condition, and the output address is correct, then a defective module is likely.

If the output module indicating LED illuminated, but a connected output device did not energize, check the following:

1. Wiring to the output device.
2. Operation of the output device.
3. Proper potential to the output device.
4. Output device wired to correct output module and proper terminal.

The second method for testing output devices, for PC's that have the option, is to use the FORCE feature. The FORCE feature, as discussed in Chapter 8, allows the user to turn an output device ON and OFF without using a pushbutton or other contacts.

Figure 17-7 shows Allen-Bradley's format for programming a rung to test output address 01000 using the FORCE ON function.

Figure 17-7. Rung for Testing Output Devices Using FORCE ON Function

The branch end instruction is used to create an open in the rung. This prevents the output from being unconditionally energized when the PC is placed in the RUN mode.

NOTE: For other PC systems, an open instruction would be used to open the rung.

By moving the cursor to the output instruction, the FORCE ON function can be initiated, which should illuminate the output module indicating LED and turn on the output device, if one is connected. If the LED does not illuminate, and the output instruction is intensified, exchange the module.

If the LED lights but the output device does not energize, proceed as described previously under testing with pushbuttons.

FINAL SYSTEM CHECK-OUT

After all input and output circuits have been tested and verified you are ready for final system check-out.

Reconnect any output loads (motors, solenoids, etc.) that were previously disconnected. In the case of motors, correct rotation needs to be established before the complete machine or process can be tested.

Using a momentary pushbutton that is part of the control panel (or using one installed specifically for this purpose) load a rung of logic as previously discussed for testing outputs (Figure 17-6) into the processor.

CAUTION: Because this part of the test will cause machine motion, make sure the machine is operational and all personnel are in the clear. Station someone at the emergency stop or disconnect location to de-energize the system if necessary.

Close the pushbutton and immediately release or open it. This momentary operation of the pushbutton is called **bumping** or **jogging,** and allows the output (motor starter) to energize only momentarily. While the motor starter was only energized momentarily, it was energized long enough to determine the direction of rotation for the motor. If the rotation is wrong, reverse any two motor leads and repeat the test for verification. Continue testing all output loads that had been previously disconnected until all function correctly.

Once all machine components are tested and correct rotation(s) are established, complete machine operation testing can be accomplished.

For final system check-out, the following steps should apply:
1. Place the processor in the program mode.
2. Clear the memory of any previous rungs used for testing.
3. Using a programming device, enter the program (ladder diagram) into memory.
4. Place the processor in the test or disable output mode, depending on the PC, and verify correctness of program.

NOTE: In the test or disable output mode the outputs cannot be energized. All logic of the circuit can be verified, input devices will function, but no outputs will come on. This step must NOT be skipped if injury to personnel or damage to equipment is to be avoided.

5. Once the circuit operation has been verified in the test or disable output mode, the processor can be placed in the run mode for final verification.
6. Make changes to the program as required (timer settings, counter presets, etc.).
7. Once the circuit is in final form and the machine or process is running correctly, it is recommended that you make a copy of the program.

TROUBLESHOOTING

The key word to effective troubleshooting is **systematic.** To be a successful troubleshooter, the technician must use a systematic approach.

A systematic approach should consist of the following steps:
1. Symptom recognition
2. Problem isolation
3. Corrective action

The technician should be aware of how the system normally functions if he or she expects to successfully troubleshoot it. When prior knowledge of system operation is not possible, the next best source of information if applicable, is the operator. Don't hesitate to ask the operator what the symptoms are and what he/she thinks the problem might be.

If no operator is available, your next best source of information is the PC system itself. While the PC can't talk, it can communicate in various ways to tell you what the problem is.

There are status lights on the processor, power supply, and I/O rack that will indicate proper operation as well as alert you to problems.

The status lights of a typical processor with built-in power supply would indicate:
1. DC POWER ON — If this LED is not lit there is a fault in the DC power supply. Check the power supply fuse and/or incoming power.
2. MODE — Indicates which operating mode the processor is in (run, halt, test, program, etc.). The fault may merely be that the keyswitch is in the wrong position.
3. PROCESSOR FAULT — When this status light is on it indicates a fault within the processor. This is a major fault and will require changing the processor module.
4. MEMORY FAULT — This status light will illuminate when a parity error exists in the transmission of data between the processor module and the memory module. Replace only one module at a time. If the first module does not correct the problem, reinstall the original module and replace the other module. If replacing the second module doesn't clear the problem then replace both modules.
5. I/O FAULT — This light would indicate a communication error between the processor and the I/O rack. First, check that communication cable(s) are fully inserted into their sockets. If available, connect a CRT-type programming device to the processor and look for error codes and/or fault messages for further diagnostic assistance.
6. STANDBY BATTERY LOW — When this LED is illuminated, the RAM back-up batteries are low and need to be replaced. While this is not a fault condition, failure to replace batteries will result in losing the program when the system is shut down or a power failure occurs.

NOTE: Refer to manufacturers' literature for explanation of error codes.

Status lights on the I/O modules will also assist you with troubleshooting problems that involve input and output devices.

If the operator tells you that the solenoid that activates a brake isn't working, the first step is to determine the address of the solenoid. Once we have the address we can use the programming device to see if the output circuit to the solenoid has been turned ON. Are the input devices closed that should be closed? Has the rest of the rung logic been completed? If the answers are yes, then we can use the output address to locate the correct rack, output module, and terminal for that address.

Output modules have LED indicators that illuminate when each of the output circuits have been turned ON. If the LED is lit for the location of the solenoid, it tells us that the problem is not with the output module, but must be with the circuit from the module to the solenoid or with the solenoid itself.

A voltage check from L2 to the terminal (as shown in Figure 17-8) will verify our conclusion that the module is OK and that the problem is either in the wire to the solenoid or the solenoid itself. Further voltage checks will locate the problem.

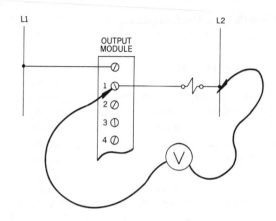

Figure 17-8. Testing Voltage on an Output Module

NOTE: On AC output modules the high internal resistance of most analog or digital meters will act like a series voltage divider when measuring across an open load (Figure 17-9). The result is a reading of nearly full voltage even after the Triac has been turned OFF. For accurate readings a 10K ohm resistor can be placed in parallel with the meter leads as shown in Figure 17-10, or a solenoid type tester (Wiggins) with low internal resistance can be used.

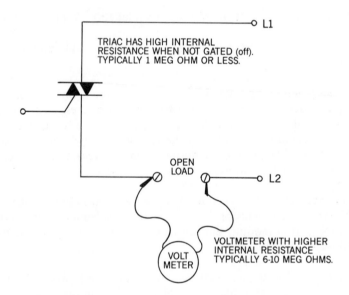

Figure 17-9. Series Voltage Divider Effect When Reading Across an Open Load

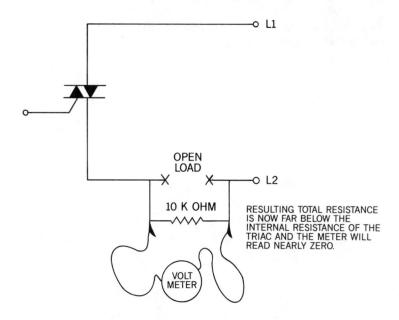

OPEN
LOAD

10 K OHM

RESULTING TOTAL RESISTANCE
IS NOW FAR BELOW THE
INTERNAL RESISTANCE OF THE
TRIAC AND THE METER WILL
READ NEARLY ZERO.

VOLT
METER

Figure 17-10. Adding a Resistor in Parallel with
Meter to Reduce Voltage Divider Effect

What would have been the procedure if the indicator LED had not been lit? The first reaction may be to change the output module, but wait! Look for a blown fuse indicator LED that will be found on most output modules. Maybe all that's needed is to replace a fuse to make the solenoid operational.

If there was no blown fuse, replacing the output module should correct our problem.

NOTE: Remove all power from the I/O rack before changing modules.

Some deluxe output modules have two indicator LEDs. One indicates that the logic from the processor has been received to turn on the output. The second LED comes on when the Triac or power transistor has been turned ON. These two LED's should come on simultaneously unless the PC is in the test mode. In test, only the logic LED will be lit because the output circuits are isolated and kept from turning ON.

Troubleshooting input modules would follow the same basic procedure. Had we determined that the solenoid had not energized because limit switch 1 was not showing closed on the programming device, we would again need to determine the address of LS-1.

From the address we can locate the hardware location where LS-1 is wired to an input module.

An illuminated LED would indicate that the limit switch was closed, but that the state of the switch (ON) was not being communicated to the processor. Exchanging the input module should make the system operational.

Had the indicator LED for LS-1 not been lit, there would be several possible problems—a bad limit switch, faulty wiring from LS-1 to the input module, or a bad input module.

Closing the limit switch and taking a voltage check as shown in Figure 17-11 will determine if the limit switch and associated wiring is OK.

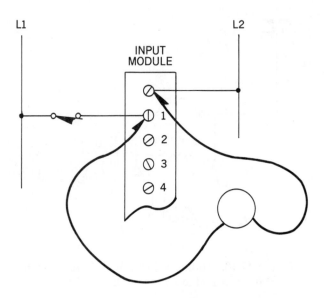

Figure 17-11. Testing Voltage on an Input Module

A voltage reading equal to the applied voltage would indicate that the limit switch and wiring are OK and that we have a faulty input module. No voltage reading indicates a problem with either the limit switch or its wiring. Further voltage checks will isolate the problem.

Like the deluxe output modules, there are deluxe input modules that have two indicating LED's. One LED indicates the input device has closed and a voltage signal has been received by the input device has closed and a voltage signal has been recived by the input module. The second LED indicates that the status of the input device (ON) has been communicated to the processor.

Chapter Summary

Start-up procedures check each part of the PC system for proper installation and operation.

Safety must always be the overriding factor when testing or operating a PC system. Care must be taken to prevent unexpected or incorrect machine motion if injury to personnel and/or damage to equipment is to be avoided.

Once the system start-up is complete and the system is operational, problems or faults can occur. To successfully troubleshoot the system, a systematic approach must be used. This systematic approach would include recognizing the symptoms, isolating the problem, and taking corrective action. A variety of indicator and status lights, as well as error messages and fault codes, will assist the technician in troubleshooting a given system.

The information covered in this chapter is intended to be general in nature and is not specific for any particular PC. For more specific information on start-up and troubleshooting procedures, refer to the manufacturer's operating manual that accompanies the PC.

Review Questions

1. A programmable controller system should be installed according to:
   a. Manufacturer's specifications
   b. Local electrical codes
   c. State electrical codes
   d. National electrical codes
   e. All of the above
2. Explain why a safety circuit or other emergency stop device is important.
3. Describe briefly how input devices are tested.
4. List two methods for testing output devices.
5. Why would output devices be disconnected before testing?
6. Draw an input module complete with input devices and power connected (L1-L2) and show how a voltage reading would be taken to verify the operation of an input device.
7. On a deluxe input module, what do the two LED indicators represent?
8. What do the two LED indicators, other than the blown fuse indicator, represent?

Appendix A

Programmable Controllers— Manufacturers, Models, & Capabilities

The information contained in this listing is as accurate as was possible to obtain. All known programmable controller manufacturers were contacted and the information in the table is a result of their returned questionnaires. Several new PC's were introduced in 1987 but the rate seems to have slowed as compared to previous years. New companies and models have been highlighted.

Figure A-1. Programmable Controllers: Manufacturers, Models, and Capabilities (1987) (Courtesy of *Control Engineering* Magazine)

Programmable Controllers

Manufacturer	Model	Total system I/O	Max discrete I/O	Max analog I/O	Relay ladder logic	High level language	PID capabilities	Motion control	Documentation	Diagnostics	Type of Interface	Scan rate/1K	Type of memory	Size of memory	Country of origin	Comments
ASEA IND'L. SYS. (New Berlin, WI)	MP51	64	64		Y			Y	Y	A		<5ms	CMOS	16K	Sweden/US	†User defined
	MP120	128	128	128	Y			Y	Y	A		†	CMOS	16K	Sweden/US	†User defined
	MP140	128	128	128	Y			Y	Y	A		†	CMOS	16K	Sweden/US	†User defined
	MP160	128	128	128	Y		Y	Y	Y	A		†	CMOS	16K	Sweden/US	†User defined
	MP150T	128	128	128	Y			Y	Y	A		†	CMOS	16K	Sweden/US	†User defined
	MP170T	128	128	128	Y			Y	Y	A		†	CMOS	16K	Sweden/US	†User defined
	MP220	5500	5500	600	Y	Y	Y	Y	Y	A		†	CMOS	1M	Sweden/US	†User defined
	MP240	1300	1300	600	Y	Y	Y	Y	Y	A		†	CMOS	1M	Sweden/US	†User defined
	MP260	1300	1300	600	Y	Y	Y	Y	Y	A		†	CMOS	1M	Sweden/US	†User defined
	MP280	1300	1300	600	Y	Y	Y	Y	Y	A		†	EEPROM, CMOS	1M	Sweden/US	†User defined
	MP200/1	5000	5000	2000	Y	Y	Y	Y	Y	B		†	CMOS	4M	Sweden	†User defined
ASC COMPUTER SYST. (St. Clair Shores, MI)	PC/86	512	512	128	Y	Y	Y	Y	Y	B		20ms	EPROM, RAM	256K	U.S.A.	AB Data Highway
	PC/88		512	64	Y	Y	Y	Y	Y	B		20ms	EPROM, RAM	1M	U.S.A.	LAN/Data Highway
	PC/188		1000	128	Y	Y	Y	Y	Y	B, C		10ms	RAM, EPROM, EEPROM	1M	U.S.A.	
ADATEK, INC. (Sandpoint, ID)	System 10	1272	1176	96	Y	Y	Y	Y	Y	A		10ms	EPROM, CMOS RAM	48K	U.S.A.	422 multidrop network
ALLEN-BRADLEY (Milwaukee, WI)	PLC-3	8192	8192	4096	Y	Y	Y	Y	Y	A		2.5ms	RAM EDC	2M	U.S.A.	†15ms/500 words
	PLC-3/10	4098	4096	2048	Y	Y	Y	Y	Y	A		2.5ms	RAM EDC	64K	U.S.A.	†2ms/500 words
	SLC 100	112	112	24	Y				Y	J		†	EEPROM, CMOS RAM	885	U.S.A./Japan	
	SLC 150	112	112	24	Y				Y	J		12.5ms	EEPROM, CMOS RAM	1200	U.S.A./Japan	
	PLC-2/02	512	256	256	Y				Y	A		12.5ms	RAM, EEPROM	1K	U.S.A.	Net. is Data Highway for all entries except SLC 100 and SLC 150
	PLC-2/16	512	256	256	Y				Y	A		12.5ms	RAM, EEPROM	3K	U.S.A.	
	PLC-2/17	512	256	256	Y				Y	A		12.5ms	RAM, EEPROM	6K	U.S.A.	
	PLC-2/30	2688	2688	896	Y				Y	A		5ms	RAM	16K	U.S.A.	
	PLC-5/12	512	512	256	Y	Y	Y	Y	Y	A		8ms	RAM, EEPROM	6K	U.S.A.	
	PLC-5/15	1024	1024	512	Y	Y	Y	Y	Y	A		8ms	RAM, EEPROM	14K	U.S.A.	
	PLC-5/25	2048	2048	1024	Y	Y	Y	Y	Y	A		8ms	RAM, EEPROM	21K	U.S.A.	
	PLC-5/VME	512	512	256	Y	Y	Y	Y	Y	†		8ms	RAM, EEPROM	14K	U.S.A.	†Backplane
ANDERSON CORNELIUS (Eden Prairie, MN)	APC-3	†	†		Y	Y	Y	Y	Y	B		18ms	PROM, RAM	80K	U.S.A.	†240/controller
	MCU-250	†	†		Y	Y	Y	Y	Y	B		40ms	PROM, RAM	80K	U.S.A.	†240/controller

Note: Codes for Type of Interface column are as follows:
A: RS-232C, B: RS-232/422, C: RS-488, D: RS-485, E: RS-449, F: 20 mA, G: 10 mA, H: Fiber optic, I: RS-423, J: RS-422, K: V24, L: TTL.

Programmable Controllers

Manufacturer / Model	Total system I/O	Max discrete I/O	Max analog I/O	Relay ladder logic	High level language	PID capabilities	Motion control	Documentation	Diagnostics	Type of interface	Scan rate/1K	Type of memory	Size of memory	Country of origin	Comments
AUTOMATIC TIMING & CONTROLS (King of Prussia, PA) ATCOM 64	72	64	8	Y	Y		Y	Y	A	7-20ms	CMOS RAM, EPROM	8K	U.S.A.	
AUTOMATION SYSTEMS (Eldridge, IA) PAC-5	1024	1024	512	Y	Y	Y	Y	Y	B	1.1ms	RAM, EPROM	32K	U.S.A.	12 commun. ports
B & R INDUSTRIAL AUTOMATION CORP. (Stone Mountain, GA) Minicontrol	96	96	8	Y	Y	Y	Y	Y	B, F	4ms	RAM, EEPROM	16K	Austria	All units use
Midicontrol	128	128	64	Y	Y	Y	Y	Y	B, F	4ms	RAM, EEPROM	16K	Austria	Ring network
Multicontrol CP40	1024	1024	128	Y	Y	Y	Y	Y	B, F	4ms	RAM, EEPROM	16K	Austria	
Multicontrol CP80	1024	1024	128	Y	Y	Y	Y	Y	B, F	2.5ms	RAM, EEPROM	74K	Austria	
PROVICON 75200	275	200	75	Y	Y	Y	Y	Y	B, F	2.5ms	RAM, EEPROM	74K	Austria	
BAILEY CONTROLS CO. (Wickliffe, OH) MPC01	1024	1024	960	Y	Y	Y		Y	A	2ms	BBRAM, ROM	276K	U.S.A.	All units use
LMM02	512	512		Y			Y	Y		15ms	BBRAM, ROM	80K	U.S.A.	proprietary network
CSC01	28	28		Y			Y	Y		2ms	BBRAM, ROM	272K	U.S.A.	
MFC03	5000	5000	5000	Y	Y		Y	Y	A	1ms	BBRAM, ROM	848K	U.S.A.	
BARBER-COLMAN CO. IND'L. INST. DIV. (Loves Park, IL) Network 8000	5000	5000	5000	Y	Y		Y	Y	A, D	5ms	RAM,EPROM,EEPROM	512K	U.S.A.	
BONAR AUGUST SYST. (Tigard, OR) Trigard/CS330S	7000	7000	3000	Y	Y		Y	Y	B	15μs	RAM, PROM	750K	U.S.A.	Uses peer-to-peer net.
BRITISH BROWN-BOVARI (Telford, England) K200	128	128		Y			Y	Y		5ms	EPROM	2K	FRG	
Procontic B	1024	768	256	Y	Y		Y	Y	B	2.5ms	RAM, EPROM	512	FRG	
DP800	4000	3000	1000		Y	Y	Y	Y	B	4ms	RAM, EPROM	inf.	UK	
P214	1164	1100†	64†	Y	Y	Y	Y	Y	B, I	2ms	RAM, EPROM	700K	UK	†per station
CINCINNATI MILACRON. ELECTR. SYST. (Lebanon, OH) APC105	64	64						Y		5ms	CMOS EEPROM	1K	FRG	
APC500 RELAY		512	64	Y		Y	Y	Y	B	5ms	CMOS RAM	8K	U.S.A.	
APC500 MCL		2048	128		Y	Y	Y	Y	B	4.5ms	CMOS RAM	8K	U.S.A.	
CONTROL SYST. INTL. (Dallas, TX) 6400	76	48	36		Y		Y	Y	D		EPROM,RAM,NOVRAM	24K	U.S.A.	Uses token passing net.
CONTROL TECHNOLOGY (Hopkinton, MA) 2200	80	80		Y		Y	Y	Y	A		BBRAM	64K	U.S.A.	
2800IEA	2208	2048	128	Y	Y	Y	Y	Y	B		BBRAM	256K	U.S.A.	Uses multidrop net.
2800IE	416	256	128	Y	Y	Y	Y	Y	B		BBRAM	256K	U.S.A.	Uses multidrop net.
2400IE	416	256	128	Y	Y	Y	Y	Y	B		BBRAM	256K	U.S.A.	Uses multidrop net.
CROUZET CONTROLS (Schaumburg, IL) CMP-31	32	32	8	Y			Y	Y		7ms	EPROM	2K	France	
CMP-34	256	256	64	Y			Y	Y	propri.	7ms	EPROM	4K	France	Uses C-line net.
CMP-340	512	512		Y			Y	Y	propri.	7ms	EPROM	8K	France	Uses C-line net.
DATEM LTD. (Ottawa, Canada) DCX538	48	48		Y	Y	Y	Y	Y	D	1ms	RAM, EPROM	64K	Canada	All units use
DCX535	19	16	3	Y	Y	Y	Y	Y	D	1ms	RAM, EPROM	64K	Canada	Bitbus network
DCX537	24	24		Y	Y	Y	Y	Y	D	1ms	RAM, EPROM	64K	Canada	
DCX536	38	32	6	Y	Y	Y	Y	Y	D	1ms	RAM, EPROM	64K	Canada	
DCX530	11	8	3	Y	Y	Y	Y	Y	D	1ms	RAM, EPROM	64K	Canada	
DCX531	22	16	6	Y	Y	Y	Y	Y	D	1ms	RAM, EPROM	64K	Canada	
DCX2000	160	†	†	Y	Y	Y	Y	Y	B, D	10ms	RAM	64K	Canada	†Expandable
DI-AN MICRO SYSTEMS (Stockport, England) DMS563	256	128	128	Y	Y	Y		Y	B	5ms	RAM	48K	UK	
DIVELBISS CORP. (Fredericktown, OH) BB-04	249	249	20	Y			Y	Y		5ms	EPROM	4K	U.S.A.	
BB-40	249	249	20	Y			Y	Y		5ms	EPROM	4K	U.S.A.	
BB-15	249	249	20	Y			Y	Y		2ms	EPROM	16K	U.S.A.	
DYNAGE CONTROLS (Cromwell, CT) SAFE 8000	2176	2048	128	Y	Y		Y	Y	B		CMOS RAM	40K	U.S.A.	
EAGLE SIGNAL CONTROLS (Austin, TX) EPTAK 100	16	16		Y			Y	Y		20ms	EEPROM	250†	Japan	†Statements
EPTAK 120	66	66		Y			Y	Y		20ms	ROM, EPROM	520†	Korea	†Statements
EPTAK 225	128	128		Y			Y	Y	B	39ms	CMOS	2800†	U.S.A.	†Statements
EPTAK 245	128	128	32	Y	Y		Y	Y	B	46ms	CMOS	2800†	U.S.A.	†Statements
EPTAK 700	2048	2048	1000		Y		Y	Y	B	1.5ms	CMOS	48K	U.S.A.	
EATON CORP. CUTLER-HAMMER (Milwaukee, WI) MPC1	128	128	8	Y				Y	A	4ms	RAM, EEPROM	2K	U.S.A.	
D100														All D100 units use
CRA28	28	28		Y			Y	Y	B	7ms	RAM, EEPROM	1K	Japan	RS-422 multi-drop
CR20	40	40		Y			Y	Y	B	7ms	RAM, EEPROM	1K	Japan	
CRA40	80	80		Y			Y	Y	B	7ms	RAM, EEPROM	1K	Japan	
CAA40H	120	120		Y			Y	Y	B	4ms	RAM, EEPROM	1K	Japan	
CRA14	34	34	2	Y			Y	Y	B	6ms	RAM, EEPROM	1K	Japan	
CR20A	40	40	2	Y			Y	Y	B	6ms	RAM, EEPROM	1K	Japan	
CRA40A	80	80	2	Y			Y	Y	B	6ms	RAM, EEPROM	1K	Japan	
D500														All D500 units use
CPU20	224	224	28	Y	Y	Y	Y	Y	B	1.5μs	RAM, EEPROM	4K	Japan	Easynet parity line
CPU25	256	256	32	Y	Y	Y	Y	Y	B	1.5μs	RAM, EEPROM	4K	Japan	
CPU50	512	512	64	Y	Y	Y	Y	Y	B	0.9μs	RAM, EEPROM	8K	Japan	
ELECTROMATIC CONTROLS CORP. (Hoffman Estates, IL) 230816	56	56		Y			Y	Y		30ms	CMOS RAM	1.5K	Denmark	
330816	56	56		Y			Y	Y		30ms	CMOS RAM	1.5K	Denmark	
300606	12	12		Y			Y	Y		30ms	CMOS RAM	1.5K	Denmark	
PLCF 223232	64	64	4	Y			Y	Y		30ms	CMOS RAM	1.5K	Denmark	Programming: built in
PLCF 323232	64	64	4	Y			Y	Y		30ms	CMOS RAM	1.5K	Denmark	Programming: external
ENCODER PROD. CO. (Sandpoint, ID) 7152	408	408	232	Y		Y	Y	Y	B		RAM, EEPROM	52K	U.S.A.	Uses BASIC
7252	264	264	88	Y		Y	Y	Y	B		EPROM, EEPROM	32K	U.S.A.	Uses BASIC
Synergy	6000	6000	2000	Y	Y	Y	Y	Y	A, D		RAM, disk	32K†	U.S.A.	†Per board

Programmable Controllers

Manufacturer	Model	Total system I/O	Max discrete I/O	Max analog I/O	Relay ladder logic	High level language	PID capabilities	Motion control	Documentation	Diagnostics	Type of interface	Scan rate/1K	Type of memory	Size of memory	Country of origin	Comments
LEHIGH FLUID POWER (Lambertville, NJ)	TPC-20	20	20		Y					Y		20ms	CMOS RAM	570	Japan	
MTS SYSTEMS CORP. (Minn., MN)	TDC/EDC	22	22			Y	Y	Y	Y	Y	A		EPROM	500	U.S.A.	Multidrop RS-485 net.
	470 SERIES	†32				Y	Y	Y	Y	Y	A		EPROM	32K	U.S.A.	†Modules
MAXITRON CORP. (Corte Madera, CA)	PLC 47Jr	80	80	16	Y	Y				Y	Y	B, F	2ms	CMOS RAM, EPROM	32K	US/France	High lev. languages:
	PLC 47-20	256	256	16	Y	Y				Y	Y	B, D, F	2ms	CMOS RAM, EPROM	32K	US/France	Grafcet & literal
	PLC 67-30	512	512	64	Y	Y	Y	Y	Y	Y	Y	B, D, F	0.45ms	CMOS RAM, EPROM	32K	US/France	for all units
	PLC 87-10	992	992	120	Y	Y	Y	Y	Y	Y	Y	B, D, F	0.45ms	CMOS RAM, EPROM	128K	US/France	
	PLC 87-30	2048	2048	256	Y	Y	Y	Y	Y	Y	Y	B, D, F	0.45ms	CMOS RAM, EPROM	128K	US/France	Maxinet One is used
	DPC 87-10	976	976	120	Y	Y	Y	Y	Y	Y	y	B, D, F	0.45ms	CMOS RAM, EPROM	2.3M	US/France	for all units
	DPC 87-30	2032	2032	124	Y	Y	Y	Y	Y	Y	Y	B, D, F	0.45ms	CMOS RAM, EPROM	4.9M	US/France	
	AC 107-30	2032	2032	124	Y	Y	Y	Y	Y	Y	Y	B, D, F	0.45ms	CMOS RAM, EPROM	17.1 M	US/France	
McGILL MFG. CO. ELEC. DIV. (Valparaiso, IN)	1701-2000	472	472	16	Y			Y	Y	Y	A	3-4ms	RAM, EPROM	4K	U.S.A.	
	1701-7000	512	512	32	Y			Y	Y	Y	A	3-4ms	RAM, EPROM	8K	U.S.A.	
MILLER FLUID POWER (Bensenville, IL)	Epic 24	48	48		Y				Y	Y		50ms	RAM, EPROM, EEPROM	640	Japan	
	Epic 40	120	120		Y				Y	Y		50ms	RAM, EPROM, EEPROM	1000	Japan	
	Epic 24K	104	104		Y				Y	Y		50ms	RAM, EPROM, EEPROM	1000	Japan	
	Epic 40K	120	120		Y				Y	Y		50ms	RAM, EPROM, EEPROM	1000	Japan	
MINARIK ELECTRIC CO. (Los Angeles, CA)	LS 1000	177	121	56	Y	Y	Y	Y	Y	Y	B	7.5ms	RAM, ROM	3K	U.S.A.	
	WP6200	12	12			Y							RAM	79	U.S.A.	
	WP6300	20	20			Y							RAM	79	U.S.A.	
MITSUBISHI ELECT. SALES AMERICA (Mt. Prospect, IL)	A1CPU	256			Y	Y	Y	Y	Y	Y	B	1.25ms	RAM, EPROM, EEPROM	6K	Japan	All units use
	A2CPU	512			Y	Y	Y	Y	Y	Y	B	1.25ms	RAM, EPROM	14K	Japan	proprietary network
	A3CPU	2048			Y	Y	Y	Y	Y	Y	B	1.25ms	RAM, EPROM	60K	Japan	
	F1-12	32	32		Y			Y	Y	Y	J	12ms	RAM, EPROM, EEPROM	1K	Japan	
	F1-20M	40	40	6	Y			Y	Y	Y	J	12ms	RAM, EPROM, EEPROM	1K	Japan	
	F1-30M	70	70	6	Y			Y	Y	Y	J	12ms	RAM, EPROM, EEPROM	1K	Japan	
	F1-40M	80	80	12	Y			Y	Y	Y	J	12ms	RAM, EPROM, EEPROM	1K	Japan	
	F1-60M	120	120	18	Y			Y	Y	Y	J	12ms	RAM, EPROM, EEPROM	1K	Japan	
	F2-40M	80	80	12	Y			Y	Y	Y	J	10ms	RAM, EPROM, EEPROM	2K	Japan	
	F2-60M	120	120	18	Y			Y	Y	Y	J	10ms	RAM, EPROM, EEPROM	2K	Japan	
NAVCOM, INC. (Huron, OH)	F10A	10	10		Y			Y	Y	Y	B, D	40ms	EPROM, RAM	32K	U.S.A.	
	F20A	20	20		Y			Y	Y	y	B, D	40ms	EPROM, RAM	32K	U.S.A.	
	F26A	26	26	16	Y			Y	Y	y	B, D	40ms	EPROM, RAM	32K	U.S.A.	
	F36B	36	36	16	Y			Y	Y	Y	B, D	40ms	EPROM, RAM	32K	U.S.A.	
	F40A	40	40		Y			Y	Y	Y	B, D	40ms	EPROM, RAM	32K	U.S.A.	
OMRON ELECTRONICS (Schaumburg, IL)	S6	64	64					Y	Y		10ms	RAM, EPROM	1024	Japan	
	C20	140	140		Y			Y	Y	H	10ms	RAM, EPROM	1194	Japan	
	C20K	84	84		Y			Y	Y	H	10ms	RAM, EPROM	1K	Japan	
	C120	256	256	22	Y			Y	Y	B, H	10ms	RAM, EPROM	2.6K	Japan	
	C500	512	512	64	Y			Y	Y	B, H	5ms	RAM, EPROM	8K	Japan	
	C200H	1024	1024		Y	Y	Y	Y	Y	B, H	0.75ms	RAM, EPROM	7K	Japan	
	C1000H	1024	1024	128	Y	Y	Y	Y	Y	B, H	0.4ms	RAM, EPROM	32K	Japan	
	C2000H	2048	2048	128	Y	Y	Y	Y	Y	B, H	0.4ms	RAM, EPROM	32K	Japan	
PHILIPS B. V. (Eindhoven, The Netherlands)	MC30	120	120		Y	Y			Y	Y	B, D	2ms	RAM, EPROM, EEPROM	2K	NL	All units use
	PC20	2000	2000	300	Y	Y	Y		Y	Y	B, D	1ms	RAM, EPROM, EEPROM	16K	NL	a proprietary network
PHOENIX DIGITAL CORP. (Phoenix, AZ)	DPAC	832	320	512	Y	Y	Y		Y	Y	B, H	Var.	CMOS RAM, UV PROM	180K	U.S.A.	Fib. optic/RS-232C nets
RELIANCE ELECTRIC (Euclid, OH)	AutoMate 15	64	64		Y			Y	Y	A	4ms	RAM, EEPROM, NVRAM	1K	U.S.A.	Uses R-NET
	AutoMate 20	256	256		Y	Y		Y	Y	A	10ms	RAM, EEPROM, NVRAM	2K	U.S.A.	Uses R-NET
	AutoMate 30	512	512	128	Y	Y		Y	Y	A	2ms	RAM, EEPROM, NVRAM	4K	U.S.A.	Uses R-NET
	AutoMate 40	8192	8192	2048	Y	Y	Y	Y	Y	A	0.8ms	RAM, EEPROM, NVRAM	104K	U.S.A.	Uses R-NET
	Shark	60	60	4	Y			Y	Y	A	5ms	EPROM, EEPROM	2K	Japan	
	Shark XL	160	160	16	Y			Y	Y	A	5ms	EPROM, EEPROM	2K	Japan	
	DCS 5000	12K+	12K+	4K+	Y	Y	Y	Y	Y	A		RAM, EPROM	80K	U.S.A.	Uses DCS NET
SELECTRON (Lyss, Switzerland)	Euroline	32	32	4				Y	Y		10ms	EPROM	1K	Switzerland	
	Picoline	28	28	4				Y	Y		10ms	EPROM	1K	Switzerland	
	128 PEX	98	98	12				Y	Y		5ms	EPROM	2K	Switzerland	
	256 PEX	112	112	12				Y	Y		5ms	EPROM	2K	Switzerland	
	PMC 20	144	144	9				Y	Y	B, D	3.5ms	RAM, EPROM, EEPROM	16K	Switzerland	Selecontrol net.
	PMC 30	256	256	128				Y		B, D	3.5ms	RAM, EPROM, EEPROM	32K	Switzerland	Selecontrol net.
SIEMENS ENERGY & AUTOMATION (Peabody, MA)	S5-100U†	128	128	8	Y					Y		70ms	RAM, EPROM, EEPROM	1K	FRG	†Uses CPU100
	S5-100U†	256	256	16	Y					D	7ms	RAM, EPROM, EEPROM	2K	FRG	†Uses CPU102
	S5-101U	64	64		Y	Y			Y		30ms	RAM, EPROM, EEPROM	1K	FRG	
	S5-101R	32	32		Y	Y			Y		2.5ms	RAM, EPROM, EEPROM	384	FRG	Most units use Sinec
	S5-105R	128	128		Y	Y			Y		5ms	RAM, EPROM, EEPROM	1000	FRG	
	S5-115U/942	2048	2048	128	Y	Y	Y	Y	Y	A	15ms	RAM, EPROM, EEPROM	42K	FRG	
	S5-115U/943	2048	2048	128	Y	Y	Y	Y	Y	A	10ms	RAM, EPROM, EEPROM	48K	FRG	
	S5-135U-R	8192	8192	384	Y	Y	Y	Y	Y	A	†	RAM, EPROM	128K	FRG	†20ms/8 loops
	S5-135U-S	8192	8192	384	Y	Y	Y	Y	Y	A	2ms	RAM, EPROM, EEPROM	128K	FRG	
	S5-150U	38K	38K	384	Y	Y	Y	Y	Y	A	1.5ms	RAM, EPROM	112K	FRG	

Programmable Controllers

Manufacturer	Model	Total system I/O	Max discrete I/O	Max analog I/O	Relay ladder logic	High level language	PID capabilities	Motion control	Documentation	Diagnostics	Type of interface	Scan rate/1K	Type of memory	Size of memory	Country of origin	Comments
SOLID CONTROLS, INC. (Minneapolis, MN)	EPIC 1	520	512	8					Y	Y	Y		2.5ms	EPROM	16K	U.S.A.	
	SYSTEM 10	136	128	8					Y	Y	Y		2.5ms	EPROM	16K	U.S.A.	
	EPIC 8B	584	384	200			Y		Y	Y	Y	B, D	1.5ms	EPROM, RAM	392K	U.S.A.	
SPRECHER & SCHUH (Aarau, Switzerland)	SESTEP 390	160	160	120	Y				Y	Y	A	20ms	EEPROM	2K	Jap./Swiss	
	SESTEP 490	2080	2080	108	Y	Y			Y	Y	B	9.6ms	RAM, EPROM	7.6K	Jap./Swiss	
	SESTEP 590	2432	2432	240	Y	Y		Y	Y	Y	B	9.6ms	RAM, EPROM	15.7K	Jap./Swiss	
	SESTEP 690	3840	3840	768	Y	Y		Y	Y	Y	B	5.1ms	RAM, EPROM	48.5K	Jap./Swiss	
	SESTEP 300	128	128						Y	Y	A	5ms	RAM, EPROM	2K	Japan	
	SESTEP 430	144	144	32	Y				Y	Y	B	50ms	RAM, EPROM	16K	Switzerland	
	SESTEP 530	1024	1024	64	Y				Y	Y	A	45ms	RAM, EPROM	16K	Switzerland	
SQUARE D CO. (Milwaukee, WI)	SY/MAX 50	256	256	32	Y					Y		7ms	RAM, EPROM, EEPROM	4K	Japan	All units use
	SY/MAX 100	40	40		Y				Y	Y	J	10ms	RAM, UVPROM	420	UK	Time Token Passing
	SY/MAX 300	256	256	112	Y	Y	Y	Y	Y	Y	J	30ms	RAM, UVPROM	2K	U.S.A.	network except
	SY/MAX 500	2000	2000	1792	Y	Y	Y	Y	Y	Y	J	2.6ms	RAM, UVPROM	8K	U.S.A.	SY/MAX 50
	SY/MAX 700	14K	14K	3584	Y	Y	Y	Y	Y	Y	J	1.3ms	RAM, Bubble	64K	U.S.A.	
	SY/MAX LC	80		28	Y	Y	Y	Y	Y	Y	J	200ms†	RAM	256	UK	†Per 4 loops
TELEMECANIQUE (Westminster, MD)	TSX 27	80	80		Y	Y			Y	Y	F	2ms	RAM, EPROM	32K	France	All units except
	MPC-007	256	256	32	Y				Y	Y	F	32ms	RAM, EPROM	4K	Japan	TSX 27 and MPC-007
	TSX 17	120	120	12	Y	Y	Y	Y	Y	Y	J	10ms	RAM, EPROM	24K	France	use peer-to-peer net.
	TSX 47 Jr	80	80	22	Y	Y	Y	Y	Y	Y	B	2ms	RAM, EPROM	32K	France	
	TSX 47	256	256	44	Y	Y	Y	Y	Y	Y	B	2ms	RAM, EPROM	32K	France	
	TSX 47-30	256	256	64	Y	Y	Y	Y	Y	Y	B	0.5ms	RAM, EPROM	32K	France	
	TSX 67-30	512	512	64	Y	Y	Y	Y	Y	Y	B	0.5ms	RAM, EPROM	32K	France	
	TSX 87-10	1024	1024	128	Y	Y	Y	Y	Y	Y	B	0.5ms	RAM, EPROM	64K	France	
	TSX 87-30	2048	2048	256	Y	Y	Y	Y	Y	Y	B	0.5ms	CMOS RAM, EPROM	128K	France	
TEMPATRON, LTD (Reading, England)	TPC 9000	252	252	60		Y	Y		Y	Y	B	10ms	RAM	32K	UK	
TENOR CO. (New Berlin, WI)	100	252	252	15	Y	Y	Y	Y	Y	Y	B, D	10ms	RAM, EPROM	32K	UK	T-NET network
	PSC 763	96	96			Y			Y	Y			EPROM	128	U.S.A.	
TEXAS INSTRUMENTS INDUSTRIAL SYST. (Johnson City, TN)	5TI	512	512		Y				Y	Y	L	8.2ms	RAM, EPROM	4K	U.S.A.	
	510	40	40		Y				Y	Y	A, L		RAM, EPROM	256	U.S.A.	All units use
	TI100	128	128		Y				Y	Y	L		RAM, EPROM	1K	Japan	TIWAY1 network
	TI160	24	18	6	Y				Y	Y	A	5ms	RAM, NOVRAM	762	U.S.A.	
	520C-1102	512	512	512	Y	Y	Y	Y	Y	Y	B, I	4ms	RAM, EPROM	3.5K	U.S.A.	
	530C-1104	1023	1023	1023	Y	Y	Y	Y	Y	Y	B, I	4ms	RAM, EPROM	8K	U.S.A.	
	530C-1108	1023	1023	1023	Y	Y	Y	Y	Y	Y	B, I	4ms	RAM, EPROM	15K	U.S.A.	
	530C-1112	1023	1023	1023	Y	Y	Y	Y	Y	Y	B, I	4ms	RAM, EPROM	20K	U.S.A.	
	525-1102	512	512	64	Y	Y	Y	Y	Y	Y	A, I	3.7ms	RAM, EPROM, EEPROM	5K	U.S.A.	
	525-1104	1023	1023	1023	Y	Y	Y	Y	Y	Y	B, I	3.7ms	RAM, EPROM, EEPROM	8K	U.S.A.	
	525-1208	1023	1023	1023	Y	Y	Y	Y	Y	Y	B, I	3.7ms	RAM, EPROM, EEPROM	15K	U.S.A.	
	525-1212	1023	1023	1023	Y	Y	Y	Y	Y	Y	B, I	3.7ms	RAM, EPROM, EEPROM	20K	U.S.A.	
	535-1204	1023	1023	1023	Y	Y	Y	Y	Y	Y	B, I	0.83ms	RAM, EPROM, EEPROM	8K	U.S.A.	
	535-1212	1023	1023	1023	Y	Y	Y	Y	Y	Y	B, I	0.83ms	RAM, EPROM, EEPROM	20K	U.S.A.	
	PM550C	640	512	128	Y		Y	Y	Y	Y	B	8ms	RAM, EPROM	7K	U.S.A.	
	560/565	8192	8192	8192	Y	Y	Y	Y	Y	Y	B, I	2.2ms	RAM/RAM EPROM	256K	U.S.A.	
	8640	46	32	14	Y	Y			Y	Y	A		RAM, EPROM	120	U.S.A.	
	8641	248	248	148	Y	Y			Y	Y	A		RAM, EPROM	256K	U.S.A.	
	8642	248	248	248	Y	Y			Y	Y	A		RAM, EPROM	256K	U.S.A.	
	8650	24	16	8	Y	Y			Y	Y	A		RAM, EPROM	128K	U.S.A.	
THESAURUS (Huntsville, AL)	CBPC-1	256	256	256	Y	Y	Y	Y	Y	Y	B, C	0.5ms	RAM	500K	U.S.A.	
	CBPC-2	512	512	512	Y	Y	Y	Y	Y	Y	B, C	0.2ms	RAM	1M	U.S.A.	
	CBPC-3	1024	1024	1024	Y	Y	Y	Y	Y	Y	B, C	0.1ms	RAM	2M	U.S.A.	
	CBPC-4	2048	2048	2048	Y	Y	Y	Y	Y	Y	B, C	0.01ms	RAM	16M	U.S.A.	
TOSHIBA (Houston, TX)	EX200	240	224	16	Y		Y	Y	Y	Y	B	9ms	CMOS RAM	4K	Japan	Tosline-30 Data Hwy
	EX250	240	256	16	Y		Y	Y	Y	Y	B	7ms	CMOS RAM	4K	Japan	Tosline-30 Data Hwy
	EX500	544	512	32	Y		Y	Y	Y	Y	B	5ms	CMOS RAM	8K	Japan	Tosline-30 Data Hwy
	EX14B	34	34		Y				Y	Y	B	60ms	CMOS RAM	1K	Japan	
	EX20-PLUS	40	40	2	Y			Y	Y	Y	B	60ms	CMOS RAM	1K	Japan	
	EX28B	28	28		Y			Y	Y	Y	B	60ms	CMOS RAM	1K	Japan	
	EX40-PLUS	80	80	2	Y			Y	Y	Y	B	60ms	CMOS RAM	1K	Japan	
TRICONEX (Irvine, CA)	TRICON	2208	2208	2208	Y	Y	Y		Y	Y	A	2.9ms	RAM, PROM	378K	U.S.A.	Modbus net.
TRIPLEX (Torrance, CA)	REGENT	2560	2560	2560	Y	Y	Y		Y	y	A	1ms	CMOS RAM	512K	U.S.A.	Modbus net.
TURNBULL CONTROLS (Reston, VA)	6433	32	32	32		Y	Y		Y	Y	J		RAM	8K	UK	ANSI X3.28 net.
UTICOR TECHNOLOGY (Bettendorf, IA)	DIR. ONE	128	128		Y			Y	Y	Y	A	20ms	RAM	1970	Japan	All units use
	DIR. 4001	384	384	128	Y		Y	Y	Y	Y	B	10ms	RAM, EEPROM	6K	U.S.A.	RS-422 net.
	DIR. 4002	64	64		Y		Y	Y	Y	Y	B	10ms	RAM, EEPROM	6K	U.S.A.	
VEEDER-ROOT CO. (Hartford, CT)	V-12	120	120	15	Y				Y	Y		40ms	CMOS RAM, EPROM	944	Japan	Standard unit
	V-12 EXP	80	80	8	Y			Y	Y	Y		45ms	CMOS RAM, EPROM	832	Japan	Expanded CPU
WESTINGHOUSE ELECTRIC CO., (Pittsburgh, PA)	PC-100	30	30							†		8ms	CMOS RAM, EPROM	320	Japan	†Special
	PC-110	112	112							†		8ms	CMOS RAM, EPROM	1K	Japan	†Special
	PC-1100	144	128	16	Y		Y	Y	Y	Y	A	8ms	CMOS RAM	3.5K	U.S.A.	All units use
	PC-900	288	256	32	Y	Y	Y	Y	Y	Y	A	20ms	CMOS RAM, EPROM	2.5K	U.S.A.	WESTNET II except
	PC-700	576	512	64	Y	Y	Y	Y	Y	Y	A	8ms	CMOS RAM	8K	U.S.A.	PC-100 & PC-110
	HPPC	8192	8192	8192	Y	Y	Y	Y	Y	Y	A	0.8ms	CMOS RAM	224K	U.S.A.	

Programmable Controllers

Manufacturer (location)	Model	Total system I/O	Max discrete I/O	Max analog I/O	Relay ladder logic	High level language	PID capabilities	Motion control	Documentation	Diagnostics	Type of interface	Scan rate/1K	Type of memory	Size of memory	Country of origin	Comments
SOLID CONTROLS, INC. (Minneapolis, MN)	EPIC 1	520	512	8			Y	Y	Y			2.5ms	EPROM	16K	U.S.A.	
	SYSTEM 10	136	128	8			Y	Y	Y			2.5ms	EPROM	16K	U.S.A.	
	EPIC 8B	584	384	200		Y	Y	Y	Y		B, D	1.5ms	EPROM, RAM	392K	U.S.A.	
SPRECHER & SCHUH (Aarau, Switzerland)	SESTEP 390	160	160	120	Y				Y	Y	A	20ms	EEPROM	2K	Jap./Swiss	
	SESTEP 490	2080	2080	108	Y	Y			Y	Y	B	9.6ms	RAM, EPROM	7.6K	Jap./Swiss	
	SESTEP 590	2432	2432	240	Y	Y		Y	Y	Y	B	9.6ms	RAM, EPROM	15.7K	Jap./Swiss	
	SESTEP 690	3840	3840	768	Y	Y		Y	Y	Y	B	5.1ms	RAM, EPROM	48.5K	Jap./Swiss	
	SESTEP 300	128	128		Y				Y	Y	A	5ms	RAM, EPROM	2K	Japan	
	SESTEP 430	144	144	32					Y	Y	A	50ms	RAM, EPROM	16K	Switzerland	
	SESTEP 530	1024	1024	64					Y	Y	A	45ms	RAM, EPROM	16K	Switzerland	
SQUARE D CO. (Milwaukee, WI)	SY/MAX 50	256	256	32	Y				Y	Y	J	7ms	RAM, EPROM, EEPROM	4K	Japan	All units use
	SY/MAX 100	40	40		Y				Y	Y	J	10ms	RAM, UVPROM	420	UK	Time Token Passing
	SY/MAX 300	256	256	112	Y	Y	Y	Y	Y	Y	J	30ms	RAM, UVPROM	2K	U.S.A.	network except
	SY/MAX 500	2000	2000	1792	Y	Y	Y	Y	Y	Y	J	2.6ms	RAM, UVPROM	8K	U.S.A.	SY/MAX 50
	SY/MAX 700	14K	14K	3584	Y	Y	Y	Y	Y	Y	J	1.3ms	RAM, Bubble	64K	U.S.A.	
	SY/MAX LC	80		28	Y	Y	Y	Y	Y	Y	J	200ms†	RAM	256	UK	†Per 4 loops
TELEMECANIQUE (Westminster, MD)	TSX 27	80	80		Y	Y				Y	F	2ms	RAM, EPROM	32K	France	All units except
	MPC-007	256	256	32	Y	Y				Y		32ms	RAM, EPROM	4K	Japan	TSX 27 and MPC-007
	TSX 17	120	120	12	Y	Y	Y	Y	Y	Y	J	10ms	RAM, EPROM	24K	France	use peer-to-peer net.
	TSX 47 Jr	80	80	22	Y	Y	Y	Y	Y	Y	B	2ms	RAM, EPROM	32K	France	
	TSX 47	256	256	44	Y	Y	Y	Y	Y	Y	B	2ms	RAM, EPROM	32K	France	
	TSX 47-30	256	256	64	Y	Y	Y	Y	Y	Y	B	0.5ms	RAM, EPROM	32K	France	
	TSX 67-30	512	512	64	Y	Y	Y	Y	Y	Y	B	0.5ms	RAM, EPROM	32K	France	
	TSX 87-10	1024	1024	128	Y	Y	Y	Y	Y	Y	B	0.5ms	RAM, EPROM	64K	France	
	TSX 87-30	2048	2048	256	Y	Y	Y	Y	Y	Y	B	0.5ms	CMOS RAM, EPROM	128K	France	
TEMPATRON, LTD (Reading, England)	TPC 9000	252	252	60		Y	Y		Y	Y	B	10ms	RAM, EPROM	32K	UK	
TENOR CO. (New Berlin, WI)	100	252	252	15	Y	Y	Y	Y	Y	Y	B, D	10ms	RAM, EPROM	32K	UK	T-NET network
	PSC 763	96	96			Y							EPROM	128	U.S.A.	
TEXAS INSTRUMENTS INDUSTRIAL SYST. (Johnson City, TN)	5TI	512	512		Y				Y	Y	L	8.2ms	RAM, EPROM	4K	U.S.A.	
	510	40	40		Y				Y	Y	A, L		RAM, EPROM	256	U.S.A.	All units use
	Ti100	128	128		Y				Y	Y	L	5ms	RAM, EPROM	1K	Japan	TIWAY1 network
	Ti180	24	18	6	Y				Y		A		RAM, NOVRAM	762	U.S.A.	
	520C-1102	512	512	512	Y	Y	Y	Y	Y	Y	B, I	4ms	RAM, EPROM	3.5K	U.S.A.	
	530C-1104	1023	1023	1023	Y	Y	Y	Y	Y	Y	B, I	4ms	RAM, EPROM	8K	U.S.A.	
	530C-1108	1023	1023	1023	Y	Y	Y	Y	Y	Y	B, I	4ms	RAM, EPROM	15K	U.S.A.	
	530C-1112	1023	1023	1023	Y	Y	Y	Y	Y	Y	B, I	4ms	RAM, EPROM	20K	U.S.A.	
	525-1102	512	512	64	Y	Y	Y	Y	Y	Y	A, I	3.7ms	RAM, EPROM, EEPROM	5K	U.S.A.	
	525-1104	1023	1023		Y	Y	Y	Y	Y	Y	B, I	3.7ms	RAM, EPROM, EEPROM	8K	U.S.A.	
	525-1208	1023	1023	1023	Y	Y	Y	Y	Y	Y	B, I	3.7ms	RAM, EPROM, EEPROM	15K	U.S.A.	
	525-1212	1023	1023	1023	Y	Y	Y	Y	Y	Y	B, I	3.7ms	RAM, EPROM, EEPROM	20K	U.S.A.	
	535-1204	1023	1023	1023	Y	Y	Y	Y	Y	Y	B, I	0.83ms	RAM, EPROM, EEPROM	8K	U.S.A.	
	535-1212	1023	1023	1023	Y	Y	Y	Y	Y	Y	B, I	0.83ms	RAM, EPROM, EEPROM	20K	U.S.A.	
	PM550C	640	512	128	Y		Y	Y	Y	Y	B	8ms	RAM, EPROM	7K	U.S.A.	
	560/565	8192	8192	8192	Y	Y	Y	Y	Y	Y	B, I	2.2ms	RAM/RAM EPROM	256K	U.S.A.	
	8640	46	32	14	Y	Y			Y	Y	A		RAM, EPROM	120	U.S.A.	
	8641	248	248	148	Y	Y			Y	Y	A		RAM, EPROM	256K	U.S.A.	
	8642	248	248	248	Y	Y			Y	Y	A		RAM, EPROM	256K	U.S.A.	
	8650	24	16	8	Y	Y			Y	Y	A		RAM, EPROM	128K	U.S.A.	
THESAURUS (Huntsville, AL)	CBPC-1	256	256	256	Y	Y	Y	Y	Y	Y	B, C	0.5ms	RAM	500K	U.S.A.	
	CBPC-2	512	512	512	Y	Y	Y	Y	Y	Y	B, C	0.2ms	RAM	1M	U.S.A.	
	CBPC-3	1024	1024	1024	Y	Y	Y	Y	Y	Y	B, C	0.1ms	RAM	2M	U.S.A.	
	CBPC-4	2048	2048	2048	Y	Y	Y	Y	Y	Y	B, C	0.01ms	RAM	16M	U.S.A.	
TOSHIBA (Houston, TX)	EX200	240	224	16	Y		Y	Y	Y	Y	B	9ms	CMOS RAM	4K	Japan	Tosline-30 Data Hwy
	EX250	240	256	16	Y		Y	Y	Y	Y	B	7ms	CMOS RAM	4K	Japan	Tosline-30 Data Hwy
	EX500	544	512	32	Y		Y	Y	Y	Y	B	5ms	CMOS RAM	8K	Japan	Tosline-30 Data Hwy
	EX14B	34	34		Y			Y	Y	Y	B	60ms	CMOS RAM	1K	Japan	
	EX20-PLUS	40	40	2	Y			Y	Y	Y	B	60ms	CMOS RAM	1K	Japan	
	EX28B	28	28		Y			Y	Y	Y	B	60ms	CMOS RAM	1K	Japan	
	EX40-PLUS	80	80	2	Y			Y	Y	Y	B	60ms	CMOS RAM	1K	Japan	
TRICONEX (Irvine, CA)	TRICON	2208	2208	2208	Y	Y	Y		Y	Y	A	2.9ms	RAM, PROM	378K	U.S.A.	Modbus net.
TRIPLEX (Torrance, CA)	REGENT	2560	2560	2560	Y	Y	Y		y	A		1ms	CMOS RAM	512K	U.S.A.	Modbus net.
TURNBULL CONTROLS (Reston, VA)	6433	32	32	32		Y	Y		Y	Y	J		RAM	8K	UK	ANSI X3.28 net.
UTICOR TECHNOLOGY (Bettendorf, IA)	DIR. ONE	128	128		Y				Y	Y	A	20ms	RAM	1970	Japan	All units use
	DIR. 4001	384	384	128	Y		Y	Y	Y	Y	B	10ms	RAM, EEPROM	6K	U.S.A.	RS-422 net.
	DIR. 4002	64	64	64	Y		Y	Y	Y	Y	B	10ms	RAM, EEPROM	6K	U.S.A.	
VEEDER-ROOT CO. (Hartford, CT)	V-12	120	120	15	Y				Y	Y		40ms	CMOS RAM, EPROM	944	Japan	Standard unit
	V-12 EXP	80	80	8	Y			Y	Y	Y		45ms	CMOS RAM, EPROM	832	Japan	Expanded CPU
WESTINGHOUSE ELECTRIC CO., (Pittsburgh, PA)	PC-100	30	30							†		8ms	CMOS RAM, EPROM	320	Japan	†Special
	PC-110	112	112							†		8ms	CMOS RAM, EPROM	1K	Japan	†Special
	PC-1100	144	128	16	Y		Y	Y	Y	Y	A	8ms	CMOS RAM	3.5K	U.S.A.	All units use
	PC-900	288	256	32	Y	Y	Y	Y	Y	Y	A	20ms	CMOS RAM, EPROM	2.5K	U.S.A.	WESTNET II except
	PC-700	576	512	64	Y	Y	Y	Y	Y	Y	A	8ms	CMOS RAM	8K	U.S.A.	PC-100 & PC-110
	HPPC	8192	8192	8192	Y	Y	Y	Y	Y	Y	A	0.8ms	CMOS RAM	224K	U.S.A.	

Appendix B
Answers to Review Questions

CHAPTER 1

1. Processor Unit
   Input/Output Section
   Programming Device
3. Discrete
5. False

CHAPTER 2

1. Houses the I/O modules, communicates with the processor.
3.

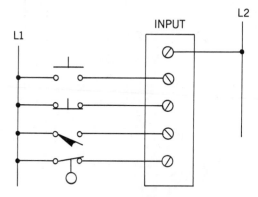

5. True
7. Handles current (loads) larger than the rating of an individual output circuit.
9. c and f
11. b
13. c

CHAPTER 3

1. Brain
3. Memory that retains stored information when power is lost or removed.
5. 8192   8 X 1024 = 8192
7. a, b, d, e, f
9. A device connected to the processor to provide an auxillary or support function.

CHAPTER 4

1. Binary
3. a. 2
   b. 16
   c. 8
5. 101100111
7. 5325
9. 350
11. a. 152
    b. 101
    c. 153
    d. 21
13. c

CHAPTER 5

1. a. S
   b. S
   c. U
   d. S
   e. U
3. a. Look at input image table to determine status of all input devices.
   b. Solve the user program.
   c. Update outputs (turns on required discrete devices).

CHAPTER 6

1. Normally open indicates the contacts are open when the device that controls them is de-energized or OFF. Normally open contacts will close when the device that controls them energizes or goes ON.
   Normally closed indicates the contacts are closed when the device that controls them is OFF and will open when the device is energized or turned ON.
3. The motor will energize when the start button is pushed but will not be sealed or maintained if 2 and 3 fail to close. When the start button is released, the motor will be de-energized.
5. a
7. a
9. Parallel
11.

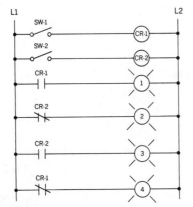

CHAPTER 7

1. A group of connected logic elements used to perform a specific function. A network is one rung of a ladder diagram. A rung can have several lines and branches but they all control the same output(s).

3. a, c

5. a. True
   b. False
   c. False
   d. True

7.

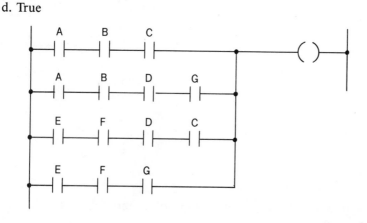

9.

CHAPTER 8

1. The programming device consists of a CRT, keyboard, and necessary circuitry for developing, modifying, and loading a program into user memory.
3. Ability to change the user program while the processor and driven equipment are running.
5. 1
7. 0
9. Indicates position on a rung.
11. False Contacts, coils, branch start and branch end instructions all use words of memory when programmed.

CHAPTER 9

1. d
3. e   Only 1 and 0 are used in the Boolean system.
5. b
7. b
9. STR 1
   STR 2
   STR NOT 3
   AND 4
   OR STR
   AND STR
   OUT 110

11.

CONTACT 1	CONTACT 2	CONTACT 3	OUTPUT 110
0	0	0	0
1	0	0	1
1	1	0	0
1	0	1	1
1	1	1	1
0	1	0	0
0	0	1	0
0	1	1	0

CHAPTER 10

1. Yes
3. So the circuit can be shut down completely without waiting for the off delay timers to time out.
5. When a device is programmed with no contacts or logic preceeding it.

CHAPTER 11

1. a. 3
   b. 2
   c. 4
   d. 1
3. False As scan time increases the accuracy is decreased.
5. b

CHAPTER 12

1. a
3. True
5. True

CHAPTER 13

1. Moving or transferring information (data) from one memory word location to another.
3. a. 4
   b. 2
   c. 6
   d. 5
   e. 7
   f. 3
5. Compare the weight of a chemical as it is placed in a mixer to the required weight needed for a given mix. When the weights are equal the chemical flow is stopped.

CHAPTER 14

1. Add, subtract, multiply and divide.
3. 1111 1100
5. 1111 1101
7. 0011 0111
9. To extend the maximum value that can be stored.

CHAPTER 15

1. Locations in memory used to store different kinds of information.
3. a
5. A group of consecutive words used to store information.
7. Transfers information from a word into a file.
9. Transfers information from one file to another file.
11. False The information is entered into the first word of the file (top) and shifts down as the instruction is needed.

CHAPTER 16

1. A sequencer is like a file to word move that acts in cylindrical or repeating fashion.
3. Determines which bit(s) of a word can be controlled by the sequencer.

CHAPTER 17

1. e
3. Manipulate device and look for indicating LED on input module.
5. To prevent damage to machinery or injury to personnel from unexpected or incorrect machine operation.
7. One indicates the input device is closed and the other indicates the status of the device (on or off) has been communicated to the processor.

GLOSSARY OF TERMS

AC INPUT MODULE: A module which converts various AC signals from "real world" (discrete) input devices to logic level for use by the processor.

AC OUTPUT MODULE: A module which converts the logic levels of the processor to an output signal to control a "real world" (discrete) output.

ADDRESS: A location in processor memory.

ANALOG INPUT MODULE: A module which converts an analog input signal to a binary or BCD number for use by the processor.

ANALOG OUTPUT MODULE: A module which provides an output proportional to a binary or a BCD number provided to the module by the processor.

ANALOG SIGNAL: A continuous signal that depends directly on magnitude (voltage or current) to represent some condition. For example, a voltage could represent the speed of a motor (e.g., 5V corresponding to 200 rpm; 10V corresponding to 400 rpm, etc.).

ARITHMETIC CAPABILITY: The ability of a PC to do addition, subtraction, multiplication and division.

ASCII: Abbreviation for American Standard Code for Information Interchange. It is a seven or eight-bit code for representing alphanumerics, punctuation marks, and certain special characters for control purposes.

BAUD: A unit of data transmission speed equal to the number of code elements (characters) per seconds.

BINARY: A numbering system that uses a base number of two. There are two digits (1 and 0) in the binary system.

BINARY CODED DECIMAL (BCD): BCD is a system of representing decimal data in binary code. For example: in BCD, 16 is represented as 0001 (for 1) and 0110 (for 6).

BIT: An acronym for **B**inary dig**IT**. A bit can assume one of two possible states; ON or OFF; high or low; logic 1 or logic 0; etc.

BOOLEAN ALGEBRA: Shorthand notation for expressing logic functions.

BOOLEAN EQUATION: Expression of relations between logic functions and/or elements.

BRANCH: A parallel logic path within a user program RUNG.

BREAKDOWN VOLTAGE: The voltage at which a disruptive discharge takes place, either through or over the surface of insulation.

BYTE: A sequence of binary digits usually operated upon as a unit. (The exact number depends on the system.)

CASCADING: A programming technique that extends the ranges of TIMER and/or COUNTER INSTRUCTIONS beyond the maximum values that may be accumulated.

CASSETTE RECORDER: A peripheral device for transferring information between PC memory and magnetic tape. In the record mode it is used to make a permanent record of a program existing in processor memory. In the playback mode it is used to enter a previously recorded program into processor memory.

CASSETTE TAPE: A magnetic recording tape permanently enclosed in protective housing. The Philips type cassette is most common. Cassettes for PC use should contain COMPUTER GRADE TAPE.

CENTRAL PROCESSING UNIT (CPU): Another term for PROCESSOR.

CHARACTER: One symbol of a set of elementary symbols, such as a letter of the alphabet or a decimal numeral.

CLOCK: A device (usually a pulse generator) that generates periodic signals for synchronization or timing.

CMOS: An abbreviation for **C**omplimentary **M**etal **O**xide **S**emiconductor. A family of very low power, high-speed integrated circuits.

CODE: A system of symbols (bits) for representing data (characters).

COMPARE FUNCTION: A program INSTRUCTION which compares numerical values for "equal," "less than," "greater than," etc.

COMPATIBILITY: The ability of various specified units to replace one another, with little or no reduction in capability.

COMPUTER GRADE TAPE: A high quality magnetic digital recording tape which must be rated at 1600 FCI (FLUX CHANGES PER INCH), or greater.

COMPUTER INTERFACE: A device designed for data communication between a central computer and another unit such as a PC processor.

CONTACT SYMBOLOGY DIAGRAM: Commonly referred to as a ladder diagram, it expresses the user-programmed logic of the controller in relay-equivalent symbols.

CORE MEMORY: A type of memory used to store information in ferrite cores. Each may be magnetized in either polarity, to represent a logical "1" or "0". This type of memory is NON-VOLATILE.

COUNTER: A device that can count up or down in response to transitions (OFF-ON) of an input signal and opens and/or closes contacts when a predetermined count is reached. Counters are internal to the processor and are not "real world" devices.

CPU: An abbreviation for **C**entral **P**rocessing **U**nit; used interchangeably with PROCESSOR.

CRT: The abbreviation for **C**athode **R**ay **T**ube, which is an electronic display tube similar to the familiar TV picture tube.

CRT TERMINAL: A terminal containing a cathode ray tube to display programs as ladder diagrams which use INSTRUCTION symbols similar to relay symbols. A CRT terminal can also display data lists and application reports.

CURSOR: A means for indicating on a CRT screen the point at which data entry or editing will occur.

DATA MANIPULATION: The process of altering and/or exchanging data between STORAGE WORDS.

DATA TRANSFER: The process of exchanging data between PC memory words and/or areas.

DIGITAL: The representation of numerical quantities by means of discrete numbers. It is possible to express in binary digital form all information stored, transferred, or processed by dual-state conditions; e.g., **on/off,** open/closed, etc. (Contrasted with ANALOG.)

DISCRETE INPUTS OR OUTPUTS: "Real world" inputs or outputs that are wired to input and output modules, as opposed to internal devices or registers.

DUMP: Recording information stored in memory onto magnetic tape or disk.

DUPLEX: A means of two way data communication. See also FULL DUPLEX and HALF DUPLEX.

ELECTRICAL-OPTICAL ISOLATOR: A device which couples input to output using a light source and detector in the same package. It is used to provide electrical isolation between input circuitry and output circuitry.

ELEMENT: A program instruction (N.O. contact, timer, counter, etc.) displayed on a CRT.

ENABLE: A circuit that allows a function or operation to be activated.

EVEN PARITY: The condition that occurs when the sum of the number of "1's" in a binary word are always even.

EXAMINE OFF: An Examine OFF PC instruction is a TRUE PRECONDITION if its addressed bit is OFF ("0"). It is FALSE if the bit is ON ("1").

EXAMINE ON: An Examine ON instruction is a TRUE PRECONDITION if its addressed bit is ON ("1"). It is FALSE if the bit is OFF ("0").

FALSE: As related to PC INSTRUCTIONS, a disabling logic state. (See TRUE.)

FAULT: Any MALFUNCTION which interferes with normal operation.

FORCE: A mode of operation or instruction on the programmer that allows the operator (as opposed to processor) to control the state of contact.

FORCE OFF FUNCTION: A feature which allows the user to de-energize, independent of the PC program, any input or output by means of the Program Panel.

FORCE ON FUNCTION: A feature which allows the user to energize, independent of the PC program, any input or output by means of the Program Panel.

FORTRAN: An acronym for **FOR**mula **TRAN**slation, a scientific programming language.

FULL DUPLEX: (FDX) A mode of communications in which data may be simultaneously transmitted and received by both ends of the circuitry.

GROUND: A conducting connection, intentional or accidental, between an electric circuit or equipment chassis and the earth ground.

GROUND POTENTIAL: Zero voltage potential with respect to earth ground.

HALF DUPLEX (HDX): A mode of data transmission capable of communicating in two directions, but in only one direction at a time.

HARD CONTACTS: Any type of physical switch contacts. Contrasted with electronic switching devices, such as triacs and transistors.

HARD COPY: Any form of printed document such as ladder diagram program listing, paper tape, or punched cards.

HARDWARE: The mechanical, electrical and electronic devices which compose a programmable controller and its application.

HARD-WIRED: Electrical devices interconnected through physical wiring.

HEXADECIMAL: The numbering system that represents all possible statuses of four bits with sixteen unique digits (0-9 then A-F).

HIGH = TRUE: A signal type wherein the higher of two voltages indicates a logic state of "1" (ON). (See LOW = TRUE.)

IEEE: Institute of Electrical and Electronics Engineers.

IMAGE TABLE: An area in PC memory dedicated to I/O data. Ones and zeroes ("1" and "0") represent ON and OFF conditions respectively. During every I/O scan, each input controls a bit in the Input Image Table; each output is controlled by a bit in the Output Image Table.

INPUT DEVICES: Devices such as limit switches, pressure switches, pushbuttons, etc., that supply data to a programmable controller. These discrete inputs are two types: those with common return, and those with individual returns (referred to as isolated inputs). Other inputs include analog devices and digital encoders.

INSTRUCTION: A command or order that will cause a PC to perform one certain prescribed operation.

INTERFACING: Interconnecting a PC with its input and output devices, and data terminals through various modules and cables. Interface modules convert PC logic levels into external signal levels, and vice versa.

I/O: An abbreviation for Input/Output. For example, a group of input modules and output modules would be referred to as I/O modules.

I/O ELECTRICAL ISOLATION: Separation of the field wiring circuits from the logic level circuits of the PC. This is typically achieved using electrical-optical isolators mounted in the I/O module.

I/O MODULE: The printed circuit assembly that interfaces between the user devices and the PC.

I/O RACK: A chassis which contains I/O Modules.

I/O SCAN TIME: The time required for the PC to monitor all inputs, read the user program and control all outputs. The I/O Scan repeats continuously.

JUMPER: A short length of conductor used to make a connection between terminals, around a break in a circuit, or around an instrument.

K: Abbreviation for 2^{10} - 1024. This abbreviation is used to denote sizes of memory, e.g., 2K - 2048.

KILO: A prefix used with units of measurment to designate quantities 1000 times as great, as in kilowatt.

LADDER DIAGRAM: A complete control scheme normally drawn as a series of contacts and coils arranged between two vertical control supply lines so that the horizontal lines of contacts appear similar to rungs of a ladder. A ladder diagram is normally the reference document used by the operator when entering the control program. (See CONTACT SYMBOLOGY DIAGRAM.)

LADDER DIAGRAM PROGRAMMING: A method of writing a user's PC program in a format similar to a relay ladder diagram.

LANGUAGE: A set of symbols and rules for representing and communicating information (data) among people, or between people and machines.

LATCH: A device that continues to store the state of the input signal after the signal is removed. The input state is stored until the latch is reset.

LATCH INSTRUCTION: A PC instruction which causes an output to stay ON, regardless of how briefly the instruction is enabled. (It can be turned OFF by an UNLATCH INSTRUCTION in a separate RUNG.)

LATCHING RELAY: A relay constructed so that it maintains a given position by mechanical means until released mechanically or electrically.

LEAST SIGNIFICANT DIGIT (LSD): The digit which represents the smallest value.

LED: Acronym for Light-Emitting Diode.

LIMIT SWITCH: A switch which is actuated by some part or motion of a machine or equipment to alter the electrical circuit associated with it.

LIQUID CRYSTAL DISPLAY: A reflective visual readout. Since its segments are displayed only by reflected light, it has extremely low power consumption - as contrasted with a LED DISPLAY which emits light.

LOAD: 1) The power delivered to a machine or apparatus. 2) A device intentionally placed in a circuit or connected to a machine or apparatus to absorb power and convert it into the desired useful form. 3) To place data (e.g., a ladder diagram) into the processor's memory.

LOGIC LEVEL: The voltage magnitude associated with signal pulses representing ones and zeroes ("1" and "0") in BINARY computation.

LOW = TRUE: A signal type where in the lower of two voltages indicates a logic state of "1" (ON). (See HIGH = TRUE.)

MAGNETIC CORE MEMORY: (See CORE MEMORY.)

MAGNETIC TAPE: Tape made of plastic and coated with magnetic material; used to store information.

MALFUNCTION: Any incorrect functioning within electronic, electrical, or mechanical hardware. (See FAULT.)

MANIPULATION: The process of controlling and monitoring bits or words by means of the user's program in order to vary application functions.

MECHANICAL DRUM PROGRAMMER: A SEQUENCER which operates switches by means of pins placed on a rotating drum. The switch sequence may be altered by changing the pin pattern.

MEMORY: A grouping of circuit elements which has data storage and retrieval capability.

MEMORY PROTECT: The hardware capability to prevent a portion of the memory from being altered by an external device. This hardware feature is under keylock control.

MILLIAMPERE (mA): One thousandth of an ampere: 10^{-3} or 0.001 ampere.

MILLISECOND (ms): One thousandth of a second: 10^{-3} or 0.001 second.

MINI-PC: A scaled-down version of a PC application with small I/O capability.

MODE: A selected method of operation, e.g., RUN, TEST, or PROGRAM.

MODEM: Acronym for **MO**odulator/**DEM**odulator. A device used to transmit and receive data by frequency-shift-keying (FSK). It converts FSK tones into their digital equivalent and vice versa.

MODULE: An interchangeable "plug-in" item containing electronic components which may be combined with other interchangeable items to form a complete unit.

MOST SIGNIFICANT DIGIT (MSD): The digit representing the greatest value.

MOTOR CONTROLLER: A device or group of devices which serves to govern, in a predetermined manner, the electrical power delivered to a motor.

NEMA STANDARDS: Consensus standards for electrical equipment approved by the majority of the members of the National Electrical Manufacturers Association (NEMA).

NETWORK: A group of connected logic elements used to perform a specific function. A network can be from one element to a complete matrix of elements, plus coil(s) as desired by the user. The size and configuration of the matrix or rung varies with the PC manufacturers.

NODE: A common connection point between two or more contacts or elements in a circuit.

NOISE: Extraneous signals; any disturbance, which causes interference with the desired signal or operation. (See INTERFERENCE.)

NOISE SPIKE: Voltage or current surge produced in the industrial operating environment.

NON-RETENTIVE OUTPUT: An output which is continuously controlled by a single program rung. Whenever the rung changes state (TRUE or FALSE), the output turns ON or OFF. (Contrasted with a RETENTIVE output which remains in its last state (ON or OFF).)

NON-VOLATILE MEMORY: A memory that is designed to retain its information while its power supply is turned off.

OCTAL NUMBERING SYSTEM: One which uses a base eight, e.g., the decimal number 324 would be written in octal notation as 504_8. Only the digits 0 through 7 are used. (See Chapter 4.)

ODD PARITY: Condition existing when the sum of the number of "1s" in a binary word is always odd.

OFF DELAY TIMER: 1) In relay-panel application, a device in which the timing period is initiated upon de-energization of its coil. 2) In PC, an instruction which counts or the delay is started whenever the rung goes FALSE.

OFFLINE PROGRAMMING: A method of programming that is done with the processor stopped and all outputs turned OFF.

ON DELAY TIMER: 1) In relay-panel applications, a device in which the timing period is initiated upon energization of its coil. 2)In PC, an INSTRUCTION which counts or the delay is started whenever the rung goes TRUE.

ON-LINE OPERATION: Operations where the programmable controller is directly controlling the machine or process.

ON-LINE PROGRAMMING: A method of programming by which rungs in the program may be inserted, changed, or deleted while the processor is running and controlling outputs under program control.

OPERAND: 1) Either of the two numbers used in a basic computation to produce an answer. For example, in the computation 2 X 3 = 6, 2 and 3 are the operands. 2) Data required for the operation of a special function.

OUTPUT: A signal provided from the Controller to the "real world". Can be either a discrete output (solenoid valve, relay, motor starter, indicator lamp, etc.), or a numerical output (e.g., display of values stored within the Controller).

OUTPUT CIRCUIT: An output module point, real-world device (e.g., motor starter, digital readout, solenoid, etc.), and its associated wiring. The output module's function is to convert processor signal levels to field voltage levels that are used by real-world devices.

OUTPUT DEVICES: Devices such as solenoids, motor starters, etc., that receive data (control) from the programmable controller.

OVERLOAD: A load greater than that which a device is designed to handle.

PARALLEL OPERATION: Type of information transfer whereby a group of digits (byte) are transmitted simultaneously.

PARITY: A method of testing the accuracy of binary numbers used in recorded, transmitted, or received data.

PARITY BIT: An additional bit added to a binary word to make the sum of the number of "1s" in a word always even or odd.

PARITY CHECK: A check that tests whether the number of "1s" in an array of binary digits is odd or even.

PC: Abbreviation for PROGRAMMABLE CONTROLLER.

PILOT DEVICE: A device used in a circuit for control apparatus that carries electrical signals for directing the performance, but does not carry the main power current.

POWER SUPPLY: In general a device which converts AC line voltage to one or more DC voltages. 1) A PC power supply provides only the DC voltages required by the electronic circuits internal to the PC. 2) A separate power supply, installed by the user, to provide any DC voltages required by the application input and output devices.

PRIORITY: Order of importance.

PROCESSOR: The part of the programmable controller that performs logic solving, program storage, and special functions within a programmable controller system and scans all the inputs and outputs in a predetermined order. The Processor monitors the status of the inputs and outputs in response to the user programmed instructions in memory, and it energizes or de-energizes outputs as a result of the logical comparisons made through these instructions.

PROGRAM: A sequence of instructions to be executed by the processor to control a machine or process.

PROGRAM PANEL (PROGRAMMER): A device for inserting, monitoring, and editing a program in a programmable controller.

PROGRAM SCAN TIME: The time required for the processor to execute all instructions in the program once.

PROGRAMMABLE CONTROLLER: A solid state control system which has a user programmable memory for storage of instructions to implement specific functions such as: I/O control logic, timing, counting, arithmetic, and data manipulation. A PC consists of the processor, input/output interface, memory, and programming device which typically uses relay-equivalent symbols. PC is purposely designed as an industrial control system which can perform functions equivalent to a relay panel or a wired solid state logic control system.

PROM: Acronym for **P**rogrammable **R**ead **O**nly **M**emory. A type of ROM that requires an electrical operation to generate the desired bit or word pattern. In use, bits or words are accessed on demand, but not changed.

PROTECTED MEMORY: Storage (memory) locations reserved for special purposes in which data cannot be entered directly by the user.

PROTOCOL: A defined means of establishing criteria for receiving and transmitting data through communication channels.

RS-232C: An Electronic Industries Association (EIA) standard for data transfer and communication.

RACK: A PC chassis that contains modules. (E.g., I/O Rack or Processor Rack.)

RAM: Acronym for **R**andom **A**ccess **M**emory. RAM is a type of memory that can be accessed (read from) or loaded (written into) depending on the particular addressing and operation codes generated internally in the PC.

RATED VOLTAGE: That maximum voltage at which an electrical component can operate for extended periods without undue degradation.

READ/WRITE MEMORY: A memory in which data can be placed (write mode) or accessed (read mode). The write mode destroys previous data; read mode does not alter stored data.

REGISTER: A location within the Controller allocated to the storage of numerical values.

REPORT: An application data display or printout containing information in a user-designed format. Reports include operator messages, part records, production lists, etc. Initially entered as MESSAGES. Reports are stored in a memory area separate from the user's program.

REPORT GENERATION: The printing or displaying of user-formatted application data by means of a programming device. Report Generation can be initiated by means of either the user's program or a programming device keyboard.

RETENTIVE OUTPUT: An output that remains in its last state (ON or OFF) depending on which of its two program rungs (one containing a LATCH INSTRUCTION, the other an UNLATCH) was the last to be TRUE. The Retentive Output remains in its last state while both rungs are false. It also remains in its last state if power is removed from, then restored to, the PC.

RETENTIVE TIMER: A PC instruction which accumulates the amount of time, whether continuous or not, that the preconditions of its rung are TRUE, and controls one or more outputs after the total accumulated time is equal to the preset time. Whenever the rung is FALSE, the accumulated time is retained. Moreover, if the outputs have been energized, they remain ON. Also, the accumulated time and energized outputs are retained if power is removed from, then restored to, the PC.

ROM: Acronym for **READ ONLY MEMORY.** A ROM is a solid state digital storage memory whose contents cannot be altered by the PC.

ROOT-MEAN-SQUARE CURRENT: The alternating value that corresponds to the direct current value that will produce the same heating effect. Abbreviated RMS.

RUNG: A grouping of PC instructions which controls one output or storage bit. This is represented as one section of a logic ladder diagram.

SCAN: The scanning operation, as performed by the processor, is the sequential examination of both the ladder diagram instructions stored in memory and the status of inputs, outputs, and registers to determine whether or not to energize or de-energize each coil, or perform the desired special functions.

SCAN TIME: The time required to make one complete scan through memory and to update the status of all inputs and outputs.

SCHEMATIC: A diagram of a circuit in which symbols illustrate circuit components.

SEQUENCER: A controller which operates an application through a fixed sequence of events. (See MECHANICAL DRUM PROGRAMMER.)

SERIAL OPERATION: Type of information transfer within a programmable controller whereby the bits are handled sequentially rather than simultaneously, as they are in parallel operation. Serial operation is slower than parallel operation for equivalent clock rate. However, only one channel is required for Serial operation.

SHIELDING: The practice of confining the electrical field around a conductor to the primary insulation of the cable by putting a conducting layer over and/or under the cable insulation. (External shielding is a conducting layer on the outside of the cable insulation. Strand or internal shielding is a conducting layer over the wire insulation.)

SOFTWARE: The user program which controls the operation of a programmable controller.

SOLID STATE: Circuitry designed using only integrated circuits, transistor, diodes, etc.; no electro-mechanical devices such as relays are utilized. High reliability is obtained with solid-state logic, which would be degraded by depending upon electro-mechanical devices.

SOLID STATE DEVICES (SEMI-CONDUCTORS): Electronic components that control electron flow through solid materials e.g., transistors, diodes, integrated circuits.

STATE: The logic "1" or "0" condition in PC memory or at a circuit's input or output.

STORAGE: Synonymous with MEMORY.

SURGE: A transient variation in the current and/or voltage at a point in the circuit.

SWITCHING: The action of turning ON and OFF a device.

SYMBOLIC NAME: A user designation for an application I/O device (E.g., S-1, LS-4, or SOL-7.)

THUMBWHEEL SWITCH: A rotating numeric switch used to input numeric information to a controller.

TIMER: In relay-panel hardware, an electro-mechanical device which can be wired and preset to control the operating interval of other devices. In PC, a timer is internal to the PROCESSOR, which is to say it is controlled by a user-programmed instruction. A timer instruction has greater capability than a hardware timer.

TOGGLE SWITCH: A panel-mounted switch with an extended lever; normally used for ON/OFF switching.

TRANSFORMER COUPLING: One method of isolating I/O devices from the controller.

TRIAC: A solid state component capable of switching alternating current.

TRUE: As related to PC INSTRUCTIONS, an enabling logic state. (See FALSE.)

TRUTH TABLE: A matrix which describes a logic function by listing all possible combinations of inputs, and by indicating the outputs for each combination.

TTL: Abbreviation for transistor/transistor logic. A family of integrated circuit logic. (Usually 5 volts is high or "1" and 0 volts is low or "0".)

UNLATCH INSTRUCTION: A PC instruction which causes an output to stay OFF, regardless of how briefly the instruction is enabled. (It can only be turned ON by a LATCH instruction in a separate RUNG.)

UV ERASABLE PROM: An erasable PROM which can be cleared (set to "0") by exposure to intense ultraviolet light. After being cleared, it may be reprogrammed. (See PROM.)

VALUE: 1) A number which represents a computed or assigned quantity. 2) A number contained in a register or FILE word.

VOLATILE MEMORY: A memory that loses its information if the power is removed from it.

WORD: A grouping or a number of bits in a sequence that is treated as a unit.

INDEX